Water and Life

and

*Life in Water
and Water in Life*

Water and Life

Life in Water and Water in Life

Arieh Ben-Naim
The Hebrew University of Jerusalem, Israel

Zvi Kirson
The Hebrew University of Jerusalem, Israel

José Angel Sordo
University of Oviedo, Spain

NEW JERSEY · LONDON · SINGAPORE · BEIJING · SHANGHAI · HONG KONG · TAIPEI · CHENNAI · TOKYO

Published by

World Scientific Publishing Co. Pte. Ltd.

5 Toh Tuck Link, Singapore 596224

USA office: 27 Warren Street, Suite 401-402, Hackensack, NJ 07601

UK office: 57 Shelton Street, Covent Garden, London WC2H 9HE

British Library Cataloguing-in-Publication Data
A catalogue record for this book is available from the British Library.

WATER AND LIFE
Life in Water and Water in Life

ISBN 978-981-122-550-5 (hardcover)
ISBN 978-981-122-628-1 (paperback)

For any available supplementary material, please visit
https://www.worldscientific.com/worldscibooks/10.1142/11972#t=suppl

Preface

Do you know on which day God created water? We asked many of our friends and colleagues this question. The majority of the people we asked hesitated at first, seemingly unsure of what they were about to say, and then provided us with answers by citing specific days (mostly the first and the second days of creation).

Others though were straightforward and readily admitted that they had no clue, or did not remember. No one among the people we asked gave the correct answer!

The reader is urged to write down his/her answer before moving on.

The subject of this book is: life in water and water in life. *Life* and *water* are both fascinating and intriguing subjects of research. Most amazingly, the Bible tells the story of the creation of life on Earth, including us humans. Surprisingly, it *does not* tell the story of the creation of *water*!

The truth of the matter is that water is mentioned in the second sentence in the book of Genesis, *before* the creation of life. Here it is:

Genesis Chapter 1:

1:1 In the *beginning* God created the Heaven and the Earth.

1:2 And the Earth was without form, and void; and darkness was upon the face of the deep. And the Spirit of God moved upon the face of the *waters*.

א בְּרֵאשִׁית, בָּרָא אֱלֹהִים, אֵת הַשָּׁמַיִם, וְאֵת הָאָרֶץ. ב וְהָאָרֶץ, הָיְתָה תֹהוּ
וָבֹהוּ, וְחֹשֶׁךְ, עַל-פְּנֵי תְהוֹם; וְרוּחַ אֱלֹהִים, מְרַחֶפֶת עַל-פְּנֵי הַמָּיִם.

Whoever wrote the story of creation must have realized that *water* preceded life, perhaps even preceding the creator of life. What an astonishing insight into the importance of water.

There are many molecules that are only found in living systems such as proteins and DNA. These are sometimes referred to as the "molecules of life". Indeed, they are. However, although water is not found exclusively in living organisms, it is nevertheless an essential and indispensable molecule for life — at least as we know it on our planet.

Life is the most exciting and mysterious phenomenon in Nature. And it is incredibly complex. The miracle of a living organism can be expressed by the question: How is it that an organism consisting of a multiplicity of tissues and cells and astronomical numbers of molecules of many different kinds can develop and function beautifully as an organized coherent whole? Equally intriguing is the understanding of the formidably complex properties of water, in terms of its marvelous molecular structure. Those two fascinating fields of research are not independent; *understanding water is an essential part in the understanding of the secrets of life.*

Water is the most studied and most unusual liquid. The very existence of water molecules in living cells is essential for the proper functioning of vital biological molecules. Even without its importance for life, liquid water is an interesting liquid in its own right.

Like in any developing scientific field, a new discovery raises new and intriguing questions to be answered and problems to be solved. We know a lot about all chemical and physical properties of pure water and aqueous solutions. We shall touch upon all these properties in this book, in a manner accessible to the lay person, to the student, to the high school teenager, and to anyone curious about scientific discoveries. We shall focus on well-known and well-understood properties, as well as touch upon controversial or incompletely understood phenomena.

The study of water by its very nature is interdisciplinary and demands harnessing the efforts of many scientists: chemists, biochemists, biologists and physicists. The reader is not required to have any previous knowledge in either of those sciences. In the first part of the book we will present in a descriptive way some of the most outstanding properties of water. In the second part we shall discuss the role played by water in some essential processes in biology.

Our primary goal is to produce as simple a book as possible without sacrificing scientific accuracy.

The book is organized into five chapters. In the first chapter we shall survey some of the outstanding properties of pure water and indicate where these properties are relevant to biology. Chapter 2 is dedicated to understanding the properties of liquid water on a molecular level. A depiction of the molecular structure will be given together with studying the nature of a chemical bond between water molecules. This information is essential for a deeper understanding of the properties of water in the liquid as well as the solid phase. In Chapters 3 and 4 we continue with the discussion of the outstanding properties of water and aqueous solutions. This is an immense field of study ranging from aqueous solutions of inert solutes, such as argon or methane, to highly complicated solutions of proteins and nucleic acids. We shall discuss only very small illustrative examples of such systems. In the last chapter we shall briefly discuss water resources and their vital role in the survival of civilizations.

We hope that this book will be informative as well as enjoyable.

Acknowledgements

Arieh Ben-Naim would like to thank his wife Roberta, for her continued help in editing and typing parts of the manuscript.

Zvi Kirson would like to dedicate this book to the memory of my father Benjamin who dedicated his life to chemistry research and education and should have enjoyed very much this book. Thanks to Arieh Ben-Naim who had been my teacher long before writing this book and paved my way into the marvels of physical chemistry. Deep thanks to my sister Rafa and her family for intensive support while being involved in working on this book.

José Ángel Sordo wishes to thank his wife, Carmen, and his sons, José Ángel and David, for their continuous support to his scientific activities. *Nenita* is always on his mind! He dedicates this book to all of them.

Contents

CHAPTER 1

Introduction: Properties and Importance of Water

Life is a unique and important phenomenon in nature. The miracle of a living organization can be stated as follows: How is it that an organism consisting of a multiplicity of diverse tissues and cells and astronomical numbers of interacting molecules of many different kinds develop and function beautifully as a whole? *Understanding water is an essential part of the answer*; possibly the first step towards understanding life, the study of which demands several branches of science such as chemistry, biochemistry and physics.

It has long been speculated that life as we know it on this planet could not have started without water. Although this is a speculation, it is a reasonable one. Strictly speaking, of course, we cannot preclude the possibility of the formation of some other kinds of life form based on other liquids.

As a fluid, water molecules interact among themselves and may combine to create tiny structures for an instant and then separate. These interactions are due to electric forces between the charged, or partly charged parts of the water molecules, leading to the so-called *hydrogen bonds* (analyzed in detail in section 2.5 of Chapter 2), on which we will elaborate later and which are responsible for most of water's unusual behavior.

Water is the major constituent of almost all life forms. It plays a wide variety of roles in life processes at different levels of complexity; from molecules and cells to tissues and organisms, it controls and regulates biochemical reactions and many life supporting processes. At the macroscopic level, water is the protagonist of highly relevant phenomena, as tides (see Appendix C).

Due to its peculiar properties, water plays an essential role in all of the cell's activities. Water has a particular effect on proteins, the building elements of the living cell. Protein–water interactions are an integral part of protein structure and stability. Water plays an important role in maintaining the specific 3D structure of proteins. It can also help protein folding and protein–protein association. Water can be bound to the cell membrane penetrating into the cell and help regulate many biochemical reactions inside the cell.

Water, as the primarily component in the blood, not only runs through our body but also runs our body. We need to constantly supply our blood with water in order to keep it going. Water as a solvent helps distribute various chemicals needed in different locations of the body and at the same time it helps to discharge other waste chemicals that are not needed, or which are even harmful to the body. But water does much more than merely transporting chemicals around the body. It affects in specific ways a multitude of biochemical processes taking place in each of our cells.

Water is the most unusual liquid in existence. No material on Earth has been investigated as thoroughly as water. Being such a familiar liquid around us, and everywhere inside us, it is difficult to believe that it is still shrouded in mystery.

The study of water offers a multitude of challenging problems to scientists, from the development of a basic understanding on the molecular level to the elucidation of its unusual properties.

Many unusual features of water are still waiting for elucidation. The deeper we probe into the inner structure of water the more enigmatic it appears to us. It should be noted that there is no general agreement or widely accepted opinions concerning the structure and function of water. We shall elaborate on this issue in later chapters. In any case, the study of water is still a lively evolving science.

Interest in water has dramatically increased over the past 50 to 60 years, especially since the discovery that water in living systems is not a "passive" liquid that *simply dissolves and transports substances from one part of an organism to another.* We now know that water plays a major part and perhaps even a decisive role in biochemical processes, and that it is vital to the maintenance of life as we know it on our planet.

Nearly all of the water on Earth is found in the oceans. The rest, only a small fraction of the total, is fresh or "sweet". Most of this fresh water is stored as ice in the Arctic and Antarctic, and some have even suggested that large icebergs might be towed from either of the poles to arid areas in order to provide a more economical source of fresh water than the desalination of ocean water.

As far as we know, water is the only liquid that supports life. It is believed that the specific anomalous properties of water are intimately related to the origin and evolution of life on Earth. Thus, we are motivated to study water because we are curious about it, and because such knowledge has many practical implications.

Water has practical implications because it is vital to our lives. A better understanding of water can lead to better medical treatments, drug design and the general improvement of the quality of life. But apart from the immense importance of water to our lives, the study of liquid water poses challenging theoretical problems unrivaled by any other liquid or substance we know. It is therefore not surprising to find that the time and effort expended in the study of water has surpassed the time and effort dedicated to the study of any other liquid.

1.1. A Few Historical Notes

We shall begin with some historical notes on the earlier theories of water and aqueous solutions, and end this chapter with some notes on the relevance of the outstanding properties of water to life. To understand *liquid* water, it is necessary to be familiar with the properties of water in the gaseous phase, as well as in the solid phase.

One can roughly divide the properties of water into two groups. The first includes properties such as the large heat capacity, the low solubility of inert solute, hydrophobic interactions, etc. On the molecular level, these were traditionally interpreted in terms of hydrogen bonds (HBs). This is basically true. However, the HBs, *per se*, are not essential to the understanding of these properties. What is essential is the *strength* of the water–water interaction. Indeed, other liquids with strong intermolecular interactions show similar behavior.

The second group consists of properties that are seemingly, as far as it is known, unique to liquid water. Examples are the negative temperature dependence of the volume, the large negative entropy of solvation of inert solute, etc. These properties were traditionally interpreted on a molecular level in terms of the *tetrahedral structure* of ice, which persists in liquid water as well.

The outstanding properties of liquid water were recognized long ago. A compilation of the properties of liquid water was first published in 1940 by Dorsey. Some of the outstanding properties of water were also discussed by Pauling (1940; 1960), Kavanau (1964), and Samoilov (1957) — the latter two are more concerned with the properties of aqueous solutions. The relevance of water to biology was discussed by Henderson (1913), Edsall and Wyman (1958) and Franks (2000). The modern era

in the research on water started in the late 1960s with the
publication of the Eisenberg and Kauzmann book in 1969. This
was followed by a book on the molecular properties of liquid
water (Ben-Naim, 1974). During the 1970s, a series of books
were published by Franks (1973–1982) that included chapters
on specific topics written by specialized authors. A more recent
book by Robinson *et al.* (1996) summarizes both the experimental and the theoretical advances in the field. An interesting, more
descriptive book was also published by Ball (1999). Finally,
two more recent books — Ben-Naim (2009; 2011) — summarize our present knowledge on the properties and theories of
water.

1.2. Properties of a Single Water Molecule

In order to understand the properties of liquid water and its
role in biological systems, one must be familiar with the basic
properties of a single water molecule.

The notation H_2O for the water molecule shows the *composition* but not the geometry of the molecule. The structure of
a single water molecule has long been established such that the
nuclei of the oxygen atom and the two hydrogen atoms form an
isosceles triangle (see Fig. 1.1).

The equilibrium O–H bond length is 0.957 Å and the H–O–
H angle is equal to 104.52°. Note that this angle is smaller than
the tetrahedral angle of 109.46°.

This particular geometry has been established by a number
of experiments utilizing a variety of methods as well as by theoretical calculations. For most of our studies of the properties
of liquid water, we can assume that a water molecule has a
rigid geometry with a fixed bond length and a fixed bond angle.
This simplified model is not sufficient for studying properties

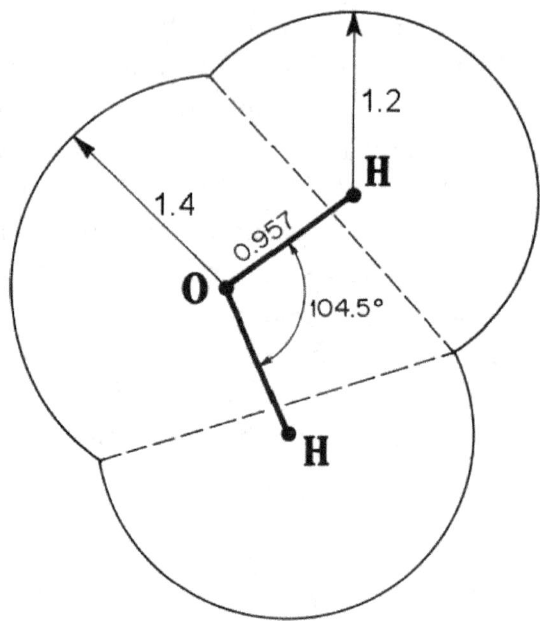

Fig. 1.1.   The geometry of a single water molecule. The van der Waals radii of the hydrogen and oxygen atoms are indicated (in Å).

that are affected by the vibrations of a single water molecule, or by the dissociation of the molecule into charged ions H^+ and OH^-.

It is important to have an idea on the order of magnitude of the energies involved in the formation of a water molecule.

The energy of formation at $0\,K$ of a molecule H_2O is defined for the process

$$H + H + O \rightarrow H_2O, \tag{1.1}$$

where H and O are the hydrogen and the oxygen atoms at the electronic ground state, and the water molecule is in the electronic, vibrational, rotational and translational ground states, i.e., we carry this hypothetical process at absolute zero temperature. The energy associated with reaction (1.1) is calculated from the energies of the following processes:

$$\text{(i) } H_2 + \frac{1}{2}O_2 \rightarrow H_2O; \quad \Delta E = -57.102\, \text{kcal mol}^{-1}$$

$$\text{(ii) } O \rightarrow \frac{1}{2}O_2; \quad\quad\quad \Delta E = -58.983\, \text{kcal mol}^{-1} \quad (1.2)$$

$$\text{(iii) } H + H \rightarrow H_2; \quad\quad \Delta E = -103.252\, \text{kcal mol}^{-1}.$$

Thus, we have:

$$\text{(i)} + \text{(ii)} + \text{(iii)} = (1.1). \quad \Delta E = -219.337\, \text{kcal mol}^{-1}. \quad (1.3)$$

Note that the *electronic binding energy* is different from the energy of formation of the water molecule, the difference between the two is the zero-point energy given in Table 1.1.

The motion of a water molecule may be described in terms of two components: (i) The motion of the center of mass of the entire molecule (this includes both translation and rotations), and (ii) the relative motions of the atoms within the molecule. The latter may be described in terms of three fundamental ("normal") modes of vibration (see Fig. 1.2).

In Table 1.1, the frequencies ν_i ($i = 1, 2, 3$) correspond to the transition from the ground state (of all the vibrational modes) to the first vibrational level of the i-th mode of vibration. The

Table 1.1. The Vibrational Frequencies and Zero Point Energies of H_2O and D_2O.[a]

	H_2O	D_2O
ν_1, symmetric stretching	3656.65	2671.46
ν_2, asymmetric stretching	3755.79	2788.05
ν_3, bending vibration	1594.59	1178.33
zero-point energy	4634.32	3388.67
	(13.2 kcal mol^{-1})	(9.8 kcal mol^{-1})

[a]From (Eisenberg and Kauzmann, 1969), based on data from Benedict *et al.* (1956) and Herzberg (1950). Values of ν_1, ν_2 and ν_3 are in cm^{-1}.

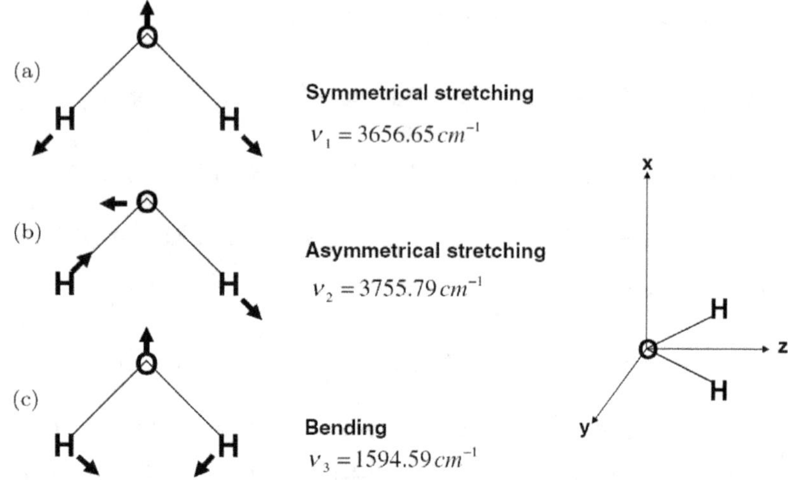

Fig. 1.2. The normal modes of vibration of a single water molecule. The first two vibrations (a) and (b) are referred to as the stretching vibrations with the corresponding frequencies ν_1 and ν_3. The third (c) is referred to as the bending vibration, with frequency ν_2.

zero-point energy of H_2O and D_2O are also given in Table 1.1. For more details, see (Eisenberg and Kauzmann, 1969).

In Table 1.2 we present some data on the molecular properties of a single water molecule. All the values in the table pertain to the hypothetical "equilibrium state" of the molecule, i.e., a molecule that does not vibrate or rotate. Note also that the bond lengths and bond angles of H_2O and D_2O are nearly the same. In studying liquid water and aqueous solutions, we can safely ignore both the internal motions (vibrations) of the atoms within molecules and the dissociation of water molecules into their ionic components. This latter dissociation phenomenon is important when we are interested in the acidity of aqueous solutions. Thus, our initial assumption is that a single water molecule has a fixed and rigid geometry. Since an oxygen atom is 16 times heavier than a hydrogen atom, the center of the mass of the oxygen atom is

Table 1.2.   Molecular Properties of H_2O and D_2O.[a]

	H_2O	D_2O
Bond length $\times 10^8$ cm	0.9572	0.9575
Bond angle	104.523°	104.474°
Moment of inertia		
$I_x \times 10^{40}$ g cm^2	2.9376	5.6698
$I_y \times 10^{40}$ g cm^2	1.9187	3.8340
$I_z \times 10^{40}$ g cm^2	1.0220	1.8384

[a] From (Eisenberg and Kauzmann, 1969) based on measurements by Benedict *et al.* (1956), all entries in this table refer to the equilibrium values. The x-axis passes through the center of mass of the molecule and is perpendicular to the plane of the molecule. The z-axis is a bisection of the bond angle and the y-axis is perpendicular to x only (see Fig. 1.2).

approximately the same as the center of mass of the entire water molecule.

The geometry of a water molecule, as depicted in Fig. 1.1, takes into account only the *nuclei* of the three atoms. Of course, there are also electrons moving around these nuclei. This "cloud" of electrons gives rise to the notion of the "volume" of the water molecule. This molecular volume is not a well-defined concept; it is not like the concept of the volume of a (macroscopic) geometrical solid such as a sphere or a cube. Since the electrons are "spread" about the three nuclei, represented as surrounding waves according to quantum mechanics, there exists no well-defined boundary to the water molecule. Nevertheless, we do assign an effective radius (hence volume) to each atom, referred to as the van der Waals radius of an atom, which is 1.4 Å for an oxygen atom and 1.2 Å for a hydrogen atom. With these van der Waals radii we can view the water molecule as though it had an *effective* solid structure. It is sometimes convenient to view a water molecule as a sphere of radius 1.41 Å. (This radius is about half of the average distance between two closest water molecules, as manifested by the first peak of the radial distribution function of water discussed later in this chapter.) We shall have more on the theoretical aspect of the chemical bonds in water in Chapter 3.

1.3. Properties of Water in the Solid Phase

Ice can have at least nine different stable and well-characterized structures. The one referred to as "ordinary ice" is the one that we are all familiar with. This one is obtained when water freezes at 0°C under ordinary atmospheric pressure. It is sometimes referred to as the *hexagonal* ice and is denoted as I_h.

The basic structure of ice was determined by Bragg in 1922 using the technique of X-ray crystallography. The "structure" of ice refers to the arrangement in space of the oxygen atoms. The structure of ordinary ice is shown in Fig. 1.3. The most important feature of the structure of ice is the *tetrahedral geometry*: Each oxygen atom is surrounded by four other oxygen atoms

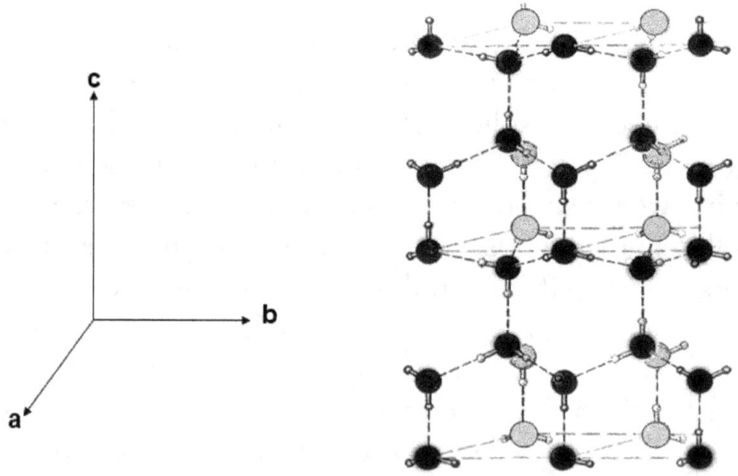

Fig. 1.3. The structure of ordinary ice I_h.

situated at the vertices of a regular tetrahedron, at a distance of 2.76 Å from the central oxygen atom.

Insofar as the geometry is concerned, the structure of D_2O is almost exactly the same as that of H_2O.

Suppose we shrink ourselves and sit at the center of an oxygen atom. We see four other oxygen atoms in our immediate vicinity, each one at a distance of 2.76 Å and all four are positioned in the four corners of a regular tetrahedron. This point of view does not reveal to us the entire 3D structure of ice I_h. Take note that along the axis labeled c in Fig. 1.3, the oxygen atoms are arranged in a structural pattern that differs from the corresponding pattern of the oxygen atoms along the two axes perpendicular to c. The difference is shown in Fig. 1.4.

If we look through an O–O nearest-neighbor bond along the c axis, we see that all the other O–O bonds are "eclipsed", i.e., on a 2D drawing they would all be superimposed. On the other hand, if we look through an O–O bond along an axis (nearly) perpendicular to the c axis, we see that three of the O–O bonds are rotated at an angle of 60° relative to the other three O–O bonds; this is the staggered configuration.

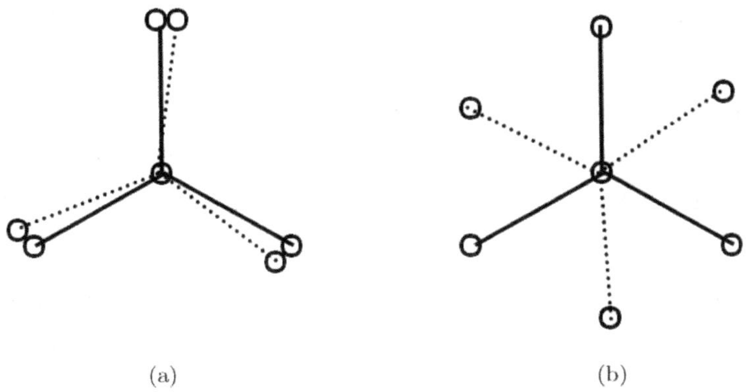

(a)   (b)

Fig. 1.4. Two views of the distribution of oxygen atoms along the O–O axis: (a) eclipsed, and (b) staggered.

This difference in structure along different axes is revealed in the way the properties of ice differ according to the measurement along different axes. For example, the coefficient of linear expansion is $63 \times 10^{-6}{}^{\circ}C^{-1}$ along the c axis but $46 \times 10^{-6}{}^{\circ}C^{-1}$ along an axis perpendicular to the c axis (at $-10{}^{\circ}C$). Similarly, the isothermal compressibility and the dielectric constant of ice are different along the different axes. For more details, see (Eisenberg and Kauzmann, 1969).

1.3.1. *The Residual Entropy of Ice*

All the structural information discussed in the previous subsections is about the locations of oxygen atoms. Using X-ray crystallography, one cannot determine the locations of the hydrogen atoms. In this section we shall discuss the *residual entropy of ice*. We shall not discuss entropy in detail here. For the purpose of this book, entropy is related to the total number of microstates, Ω, by the Boltzmann equation (see Eq. (1.6) below).

In 1933, Bernal and Fowler concluded from available experimental data that water molecules in ice maintain their molecular identity, and that the oxygen atoms are located at the corners of regular tetrahedron, the center of which is occupied by an oxygen atom. They formulated the so-called two ice conditions (see Fig. 1.5).

 (i) Each O–O line accommodates only one hydrogen atom.
(ii) Each oxygen atom has two hydrogen atoms that are at a distance of about 1 Å and two hydrogen atoms at a distance of about 1.76 Å.

These two ice conditions reflect the experimental findings that when ice is formed from either the gas or the liquid phase, water retains its identity as a single molecule, i.e., one can identify single water molecules in the ice lattice. In addition, each pair of the nearest-neighbor water molecules forms hydrogen

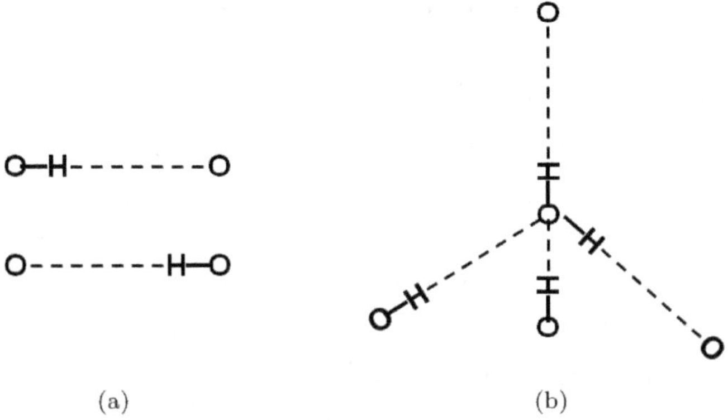

Fig. 1.5.   The two ice conditions: (a) fulfilling condition (i), and (b) fulfilling condition (ii).

bonds in such a way that only *one* hydrogen atom is situated along the O–O line (see Fig. 1.5).

In this discussion of the structure of ice, we are disregarding the very small concentrations of ions such as OH^- and H_3O^+. Such local defects do not conform to the ice conditions. Given the structural arrangement of the oxygen atoms and assuming the validity of the two ice conditions, there is still a large number of possible arrangements for the hydrogen atoms.

In 1935, Pauling calculated the approximate number of possible configurations for the hydrogen atoms that are consistent with the ice conditions. This calculation is now considered a classical example of a successful prediction based on an elementary probabilistic argument. Because of its historical importance and its didactic simplicity, we present Pauling's solution to this problem here.

Consider a perfect ice structure containing N water molecules, i.e., N oxygen atoms are situated at N lattice points. The question is: In how many ways can we arrange the $2N$ hydrogen atoms on this lattice such that the two ice conditions are fulfilled?

Since we are dealing with a macroscopic quantity of ice, N is of the order of an Avogadro number. We can neglect surface effects in our calculations.

For a very large crystal, N oxygen atoms produce $2N$ nearest-neighbor bonds O–O. We need to place the $2N$ hydrogen atoms in such a way that the two ice conditions are fulfilled. We first position the hydrogen atoms on the O–O bonds so that one hydrogen atom falls on each of the O–O bonds. Each O–O bond provides two locations for placing the hydrogen atom. These we denote O–H$\cdots$O and O$\cdots$H–O; in the first placement, the H is near the oxygen atom on the left hand side, and in the second, it is near the oxygen atom on the right hand side. See Fig. 1.5(a).

Since each of the hydrogen atoms can be placed at one of the two locations on the O–O bond, there are a total of 2^{2N} configurations that fulfill condition (i). Clearly, not all of the 2^{2N} configurations fulfill condition (ii). In Fig. 1.6, we show a system of five oxygen atoms and all of the possible arrangements of hydrogen atoms around it. Most of these configurations are not consistent with condition (ii). The next question is: How many of the 2^{2N} configurations are consistent with the second ice condition?

If we focus on a single oxygen atom and its four oxygen neighbors, we can make the following list of configuration types (see Fig. 1.6, from left to right).

Fig. 1.6.   The five possible arrangements of hydrogen atoms arounda given oxygen atom.

1. One arrangement in which all hydrogen atoms are close to the central oxygen atom.
2. Four arrangements in which one hydrogen atom is close to the central oxygen atom, while the other three are far away.
3. Six arrangements in which two hydrogen atoms are close to the central oxygen atom, while the other two are far away.
4. Four arrangements in which three hydrogen atoms are close to the central oxygen atom, while the other one is far away.
5. One arrangement in which all hydrogen atoms are far away from the central oxygen atom.

Altogether, there are 16 different arrangements divided into 5 groups.

$$\text{Group:}\quad 1\quad 2\quad 3\quad 4\quad 5$$
$$\text{Number of arrangements:}\quad 1\quad 4\quad 6\quad 4\quad 1$$

Of these 16 arrangements, all of which are consistent with the first ice condition, only 6 arrangements (i.e., those in Group 3) are also consistent with the second ice condition.

Pauling's reasoning is as follows: First assume that the fraction 6/16 is also the *probability* of finding a single oxygen atom with the right configuration (i.e., one that fulfills the second ice condition), then assume that all oxygen atoms are independent. The probability that all oxygen atoms have configurations that fulfill the second condition is therefore $(6/16)^N$. This is also the fraction of the total configurations that fulfill the two ice conditions. Hence, the number of the configurations that fulfill the two ice conditions is

$$\Omega = \left(\frac{6}{16}\right)^N 2^{2N} = \left(\frac{3}{2}\right)^N = 1.5^N. \tag{1.4}$$

This result would not have been so remarkable had it not been related to the residual entropy of ice. For details on the entropy and the residual entropy of ice see Ben-Naim (2009, 2016).

The experimental value of the residual entropy of ice is known

$$S_{\text{exp}} = 0.81\,\text{cal}\,\text{mol}^{-1}\,\text{K}^{-1}. \tag{1.5}$$

Assuming that the residual entropy is determined by the number of configurations Ω in Eq. (1.4), we get the theoretical value of the residual entropy:

$$S_{th} = k_B \ln \Omega = k_B \ln(1.5)^N = k_B \ln 1.5 = 0.805\,\text{cal}\,\text{mol}^{-1}\,\text{K}^{-1}. \tag{1.6}$$

The relation between S and Ω, as shown in Eq. (1.6), is the famous Boltzmann equation. It is engraved on Boltzmann's tombstone, which can be found in Vienna's Zentralfriedhof cemetery (see Fig. 1.7).

Fig. 1.7.   Boltzmann's tombstone in Vienna's Zentralfriedhof cemetery.

The formula $S = k \log W$ may be seen, rather fuzzy, at the top of the figure.

This is a very remarkable relation between a thermodynamic quantity (the entropy) on one hand, and the number of configurations (micro-states) of a system on the other hand. The larger the number of configurations (having equal probabilities), the larger the uncertainty, or the larger the missing information on the system. This is the essence of the meaning of entropy. For more details, see (Ben-Naim, 2009; 2016). It is also a remarkable achievement that in spite of the drastic approximations made in calculating the number of configurations in Eq. (1.4), the experimental value of residual entropy of ice agrees quite well with the theoretical value, based on Pauling's calculation. The agreement between the experimental and the theoretical values of the residual entropy of ice clearly indicates that the distribution of hydrogen atoms within the ice lattice is not unique. There are many possible configurations that are consistent with the two ice conditions.

Note that to fulfill the second ice condition, we must maintain water molecules in ice as single entities, i.e., a pair of hydrogen atoms must "belong" to each oxygen atom. This does not mean that the structure of a single water molecule is the same as that of molecules in the gas phase. We recall that the H–O–H angle in a single water molecule in the gas phase is 104.5°. In a perfect ice crystal, the tetrahedral angle (a characteristic of a triplet of neighboring oxygen atoms O–O–O) is 109.5°, definitely larger than 104.5°.

1.4. The Phase Diagram of Water

Figure 1.8 shows the phase diagram of water at low pressures (about one atmosphere). There are three regions denoted as: Solid (I_h), liquid and vapor. In each of these regions, one can

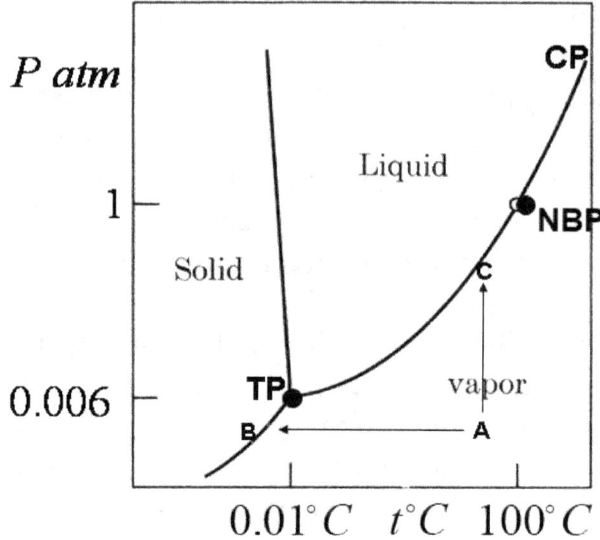

Fig. 1.8. Phase diagram of water. The triple point (TP), the critical point (CP), and the normal boiling point (NBP) are indicated.

change both the pressure P and the temperature T, and still *observe only a single phase* (see section 3.6.2 in Chapter 3). For instance, suppose we start at point A in the figure with chosen values of temperature T and pressure P, corresponding to a *system of pure vapor*. In this state, we have *two degrees of freedom*, i.e., we can change both the pressure P and the temperature T, and still observe a single pure phase. We can wander in each of these areas, i.e., changing both P and T independently without encountering a new phase. However, when we reach one of the boundary lines, a new phase appears. For example, suppose we start at point A. As long as we make small changes in either P or T, we stay in the vapor area. If we keep the pressure fixed and decrease the temperature, we shall be moving along the line AB. As long as we do not hit the boundary line between vapor and solid, we have pure vapor. Once we reach point B, ice appears. The system now consists of two phases at equilibrium. The phase rule of thermodynamics tells us that a system of two phases at equilibrium has only *one degree of freedom*. This means that we cannot change both P and T independently. Changing for instance the temperature determines the pressure, i.e., we have the function $P = f_{VS}(T)$, which is the boundary curve between the vapor and the solid phases. Clearly, in this diagram, we have three such functions: One for the vapor–liquid boundary, one for the vapor–solid boundary, and one for the liquid–solid boundary. Along each of these curves are two phases at equilibrium. As long as we change the temperature as well as the pressure in such a way that we "move" along one of these lines, we maintain the two phases at equilibrium.

Starting again at point A, keeping the temperature constant and increasing the pressure, we move along the AC line upward. As long as we do not hit point C, we observe one phase only — pure vapor. At point C, liquid appears and we observe two phases — liquid and vapor at equilibrium. Again, when we move along this line, while maintaining the two phases, we must follow the corresponding function $P = f_{VL}(T)$, which describes the dependence between the two degrees of freedom P and T. Note that the three curves in this phase diagram are described by three different functions. We denote these functions by $f_{VL}(T), f_{VS}(T)$, and $f_{SL}(T)$ for the vapor–liquid, vapor–solid and solid–liquid phases, respectively.

Moving along any of the boundary lines, e.g., the vapor–solid line, we observe two phases at equilibrium. This is true until we reach a point, denoted TP in the diagram, at which a third phase appears. We now have three phases — vapor, liquid and solid — at equilibrium. At this point, the phase rule of thermodynamics tells us that there are *zero degrees of freedom*. In other words, we cannot "move" in the phase diagram while observing the three phases at equilibrium. The point at which the three phases exist at equilibrium is called the triple point (TP).

This is a unique point in the phase diagram and is characterized by $P = 0.006$ atm and $t = 0.01°C.$[1]

Note that since three phases are at equilibrium at the triple point, it must be the intersection of the three boundary lines. In other words, at the triple point, we have the equalities:

$$P_{TP} = f_{VL}(T_{TP}) = f_{VS}(T_{TP}) = f_{SL}(T_{TP}). \qquad (1.7)$$

In Fig. 1.8, we noted one more unique point denoted by CP. This is the vapor–liquid critical point. When we increase the temperature but follow the vapor–liquid boundary line, i.e., when we move along the curve $P = f_{VL}(T)$, we eventually reach a point, referred to as the critical point, where there is no distinction between the vapor and the liquid phases. The two phases become one. This point is characterized by the pressure $P_{CP} = 218$ atm and $T_{CP} = 374.15°C$. The molar volume of the water at the critical point is $59.1 \text{ cm}^3 \text{ mol}^{-1}$.

Note that both the triple and critical points are uniquely defined in the phase diagram. They are fundamentally different kinds of points. The triple point, here of vapor–liquid–ice I_h, is characterized by the coexistence of three phases at equilibrium. On the other hand, when we approach the critical point along the vapor–liquid boundary line, the two phases become more and more similar. Similar in the sense that the densities of the two phases become closer and closer. At the critical point, the densities of the vapor and the liquid phases become identical, hence, we observe only a single phase. Beyond the critical point, i.e., increasing either the pressure, or the temperature, there exists only one phase that is referred to as a fluid. The fluid may be viewed as either a highly compressed gas, or as an expanded liquid.

If we move along the solid–liquid boundary, do we encounter a new critical point? The answer is "no". We might encounter

---

[1] $1 \text{ kbar} = 10^9 \text{ dyn cm}^{-2} = 986.9 \text{ atm}.$

other triple points (see below) but we do not expect to find another critical point. The reason is that there exists a fundamental difference between a solid phase (any solid, not necessarily ice), and either a liquid or gaseous phase. The liquid and the gas phases are randomly disordered systems; the two phases differ in their densities. When we move along the vapor–liquid curve, the difference in the densities of the two phases become smaller and smaller until it disappears at the critical point. On the other hand, a solid is an ordered phase. When it is at equilibrium with either a gas or a liquid, there is a clear-cut distinction between the two phases. Order and disorder cannot gradually change until they are equal.

Note that the slopes of the liquid–vapor and the solid–vapor are always positive. This is understandable; changing from a condensed phase (solid or liquid) to the vapor involves an increase in entropy (or equivalently of enthalpy) and an increase in volume.

In most substances the solid–liquid curve has a positive slope too, for the same reason given above. Water is anomalous and as can be seen from Fig. 1.8, the slope of the solid–liquid curve is negative. This is an important observation. We shall discuss its molecular implication in the next chapter.

Most substances have a phase diagram similar to that of Fig. 1.8. They differ in their triple-point and critical point locations, and of course in the location of the boundaries between phase pairs.

1.5. Properties of Water in the Liquid Phase

In this section, we shall survey some of the outstanding properties of liquid water. We shall discuss only equilibrium thermodynamic quantities. Sometimes these properties are referred to as *anomalous* properties or *unique* properties. We shall occasionally use these terms as well as the term *outstanding* in describing those properties of liquid water that differ considerably from the values of the same properties in other liquids. To claim uniqueness is not always justified unless one examines the properties of *all* other liquids and finds that water has indeed a unique property.

Consider the following series of substances: Methane (CH_4), ammonia (NH_3), water (H_2O), hydrofluoric acid (HF), and neon (Ne). These substances are all isoelectronic, i.e., they all have the same number of electrons each — 10 (hence also 10 protons, since the molecules themselves are electro-neutral).

Table 1.3 shows the values of the melting point (in °C), the boiling point (in °C) and the heat of evaporation (in cal mol^{-1}) of each of these substances. Looking at each of these three columns we notice that the values first increase, then sharply decrease, and that the maximal value is attained by liquid water. The same information is also shown graphically in Fig. 1.9, where, in addition to the series of substances reported in Table 1.3, we have a series of compounds of the general formula RH_n, where R is a varying atom along a column of the periodic table of the elements and n is the number of hydrogen (H) atoms in the compound.

Table 1.3.   Some Physical Properties of a Series of Isoelectronic Substances.

Substance	Melting Point	Boiling Point	Molar Heat of Vaporization (cal mol^{-1})
CH_4	−184	−161.5	2200
NH_3	−78	−33.4	5550
H_2O	0	+100	9750
HF	−92	+19.4	7220
Ne	−249	−246	415

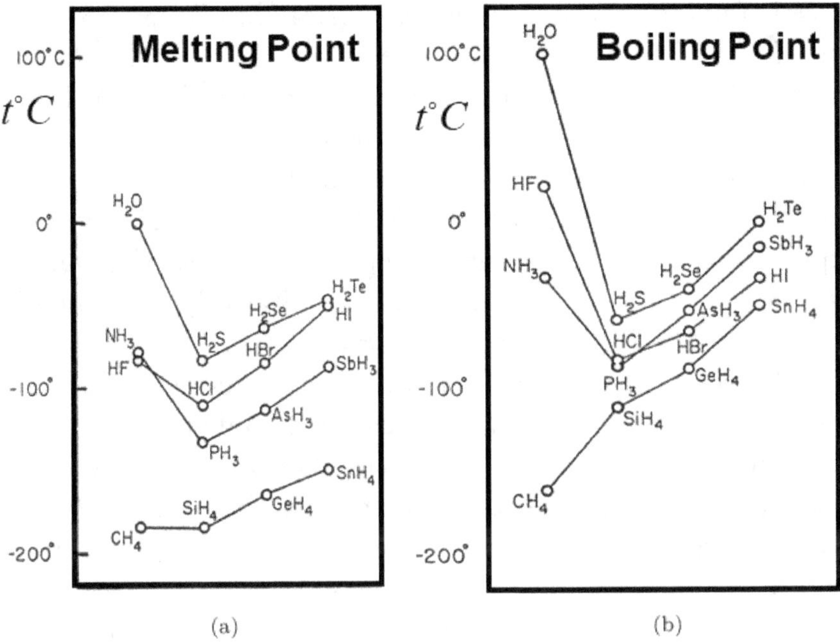

Fig. 1.9. (a) The melting points, and (b) boiling points of isoelectronic sequences of hydride molecules.

Consider first the series of CH_4, SiH_4, GeH_4 and SnH_4 (see Fig. 1.9). We see that in this homologous series, the properties change "regularly" with the molecular weight, or with the size of the molecule. Note that the ordinates in these figures report the numerical values of the melting points and boiling points. The abscissa, on the other hand, indicates only the row in the periodic table from which the element R, in RH_4, has been taken. From this series, one can conclude that the larger the molecular weight of the substance, the higher the melting and boiling temperatures.

We see clearly that the properties of the substances HF, H_2O and NH_3 have high values — far above the values that we would have derived if we extrapolated from the values of the same property of the homologous substance. This is most conspicuous for the homologous substances H_2O, H_2S, H_2Se and H_2Te. Although the last three compounds roughly form a straight line with increasing values from H_2S to H_2Se to H_2Te, the value of water is outstandingly high.

The high values of the melting points, the boiling points, and the heat of vaporizations of H_2O, NH_3 and HF can all be explained by introducing the concept of hydrogen bonding (see section 2.5 in Chapter 2). This was qualitatively explained by Pauling in *The Nature of the Chemical Bond*, first published in 1939. A hydrogen bond is not a chemical bond in the usual sense. Although a hydrogen atom can form only one pure covalent bond, a hydrogen *ion* H^+ — with its small radius — exerts a strong electrostatic force that attracts simultaneously two electro-negative atoms such as oxygen atoms (electro-negativity is the tendency of an atom to attract a bonding pair of electrons).

Figure 1.10 shows how the small proton H^+ can bring two relatively large negative ions (denoted as A_1^- and A_2^-) in close proximity. As can be seen in Fig. 1.10, a third ion (denoted as A_3^-) approaching this pair cannot come close to the proton. Hence effectively, the hydrogen ions act as a "glue" that binds no

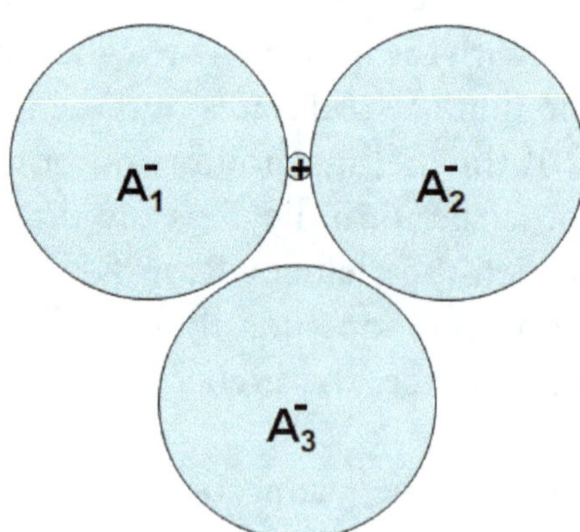

Fig. 1.10.   Schematic description of the electrostatic interaction in hydrogen-bonding.

more than two negative ions together. This type of "saturation" is typical in a chemical bond and can be observed in all hydrogen bonds.

It has also been found that the greater the electro-negativity of the ions, the stronger the hydrogen bonds. Fluoride ions form very strong hydrogen bonds, oxygen forms somewhat weaker hydrogen bonds, nitrogen forms still weaker hydrogen bonds, and carbon almost does not form hydrogen bonds.

Note that although the strength of a single hydrogen bond increases from nitrogen to oxygen to fluoride, the effect of hydrogen-bonding is more dramatic in water than in hydrofluoric acid. The reason is that in liquid water, a water molecule can form up to *four* hydrogen bonds with its neighboring molecules.

Viewing the phenomenon of hydrogen-bonding as a force of adhesion between water molecules in the liquid state, we can immediately understand the high values of the melting point, boiling point and heat vaporization. Since the forces that hold water molecules together are strong, a relatively high temperature is needed to melt ice. Note that we are comparing the melting point of pure water with the melting points of only those substances shown in Fig. 1.9.

There are many other substances with much stronger binding forces, hence higher melting points. Table 1.4 shows the values of melting and boiling points of H_2O, D_2O and T_2O. The larger values of both the melting points and the boiling points of D_2O

Table 1.4. **Melting and Boiling Points for Water and Heavy Water.**

	Melting Point	Boiling Point
H_2O	0.00°C	100.0°C
D_2O	3.82°C	101.42°C
T_2O	4.49°C	101.51°C

and T_2O indicate stronger interaction energies among D_2O and T_2O molecules compared with H_2O.

Throughout the entire liquid range of temperatures — 0°C to 100°C — the average number of hydrogen bonds decreases as the temperature increases. However, even at the boiling point, there are many water molecules still engaged in hydrogen bonding. To detach water molecules from the liquid and to send them into the gaseous phase, one needs more energy, which explains the relatively higher boiling temperature of water and high value of the heat vaporization.

1.5.1. *Molar Volume of Water and Its Temperature Dependence*

Among the other unusual properties of pure liquid water, one of the best known is the manner in which water expands when it freezes.[2] Never put a bottle filled with water inside the freezer. It will explode upon freezing. This is quite an unusual phenomenon (though not entirely unique to water — there are other substances that also have a larger molar volume in the solid state compared to the liquid state). The molecular reason for this phenomenon is clear. It is due to the "open" mode of packing water molecules in ordinary ice.

An even more unusual property, one which is unique to liquid water is the continual decrease of the molar volume of water upon increasing the temperature between 0°C to 4°C (for heavy water (D_2O) up to 11°C) (see Fig. 1.11).

Figure 1.11(a) shows the temperature dependence of the volume of water and heavy water, and also compares methanol and ethanol (see Fig. 1.11(c)).

All known liquids expand upon temperature increase, whereas liquid water between 0°C and 4°C show the opposite

---

[2]The molar volume of water and ice at the normal freezing points are 18 and 19.6 $cm^3\, mol^{-1}$, respectively.

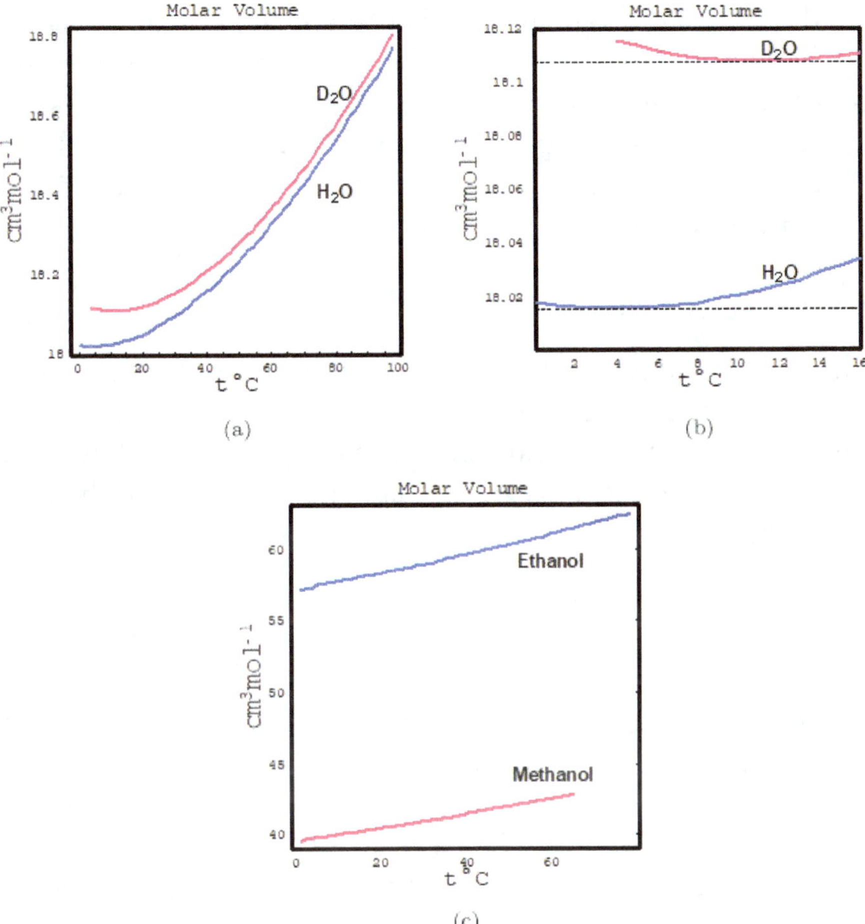

Fig. 1.11. The temperature dependence of the molar volume of (a) water and heavy water, (b) a magnified figure near the minimum of the volume, and (c) methanol and ethanol.

behavior. Note also that D_2O has a larger molar volume in the entire range of temperatures. The locations of the minimum volume are indicated in Fig. 1.11(b). This phenomenon is also due to the mode of packing water molecules in such a way that low local density is correlated with strong binding energy. We shall discuss this principle in detail later in this book. We shall see that the understanding of this phenomenon on a molecular level is essential to the understanding of some of the outstanding properties of aqueous solutions.

As we increase the pressure, the minimum of the volume is shifted to lower temperatures. At higher pressures the minimum disappears and water behaves as a normal liquid.

1.5.2. *Heat Capacity*

The heat capacity is defined as the amount of heat required to raise the temperature of one mole of a substance by one degree. The value of the heat capacity depends on the conditions in which the experiment is carried out. The most commonly used heat capacities are:

$$C_V = \left(\frac{\partial E}{\partial T}\right)_V, \quad C_p = \left(\frac{\partial H}{\partial T}\right)_P, \tag{1.8}$$

where E is the internal energy of the system and $H (H = E + PV)$ is the enthalpy of the system.

The heat capacity of water is much larger than that of other normal liquids. For example, the heat capacity of water at $25°C$ is about $1 \, \text{cal} \, \text{g}^{-1} \, \text{K}^{-1}$. For ethanol and hexane, their heat capacities are both $0.6 \, \text{cal} \, \text{g}^{-1} \, \text{K}^{-1}$, but for ammonia, its heat capacity is $1.23 \, \text{cal} \, \text{g}^{-1} \, \text{K}^{-1}$. When we say large, we mean larger than the value that is expected from a regular non-hydrogen bonded fluid. Note that in both the solid and the gaseous phases, the value of the heat capacity is "normal", i.e., it is consistent with what one would calculate from statistical thermodynamics. In the liquid phase, however, the value of the heat capacity is almost three times larger than expected.

Figure 1.12 shows the values of the heat capacity C_P as a function of T for H_2O, D_2O, and for comparison also for ethanol and methanol.

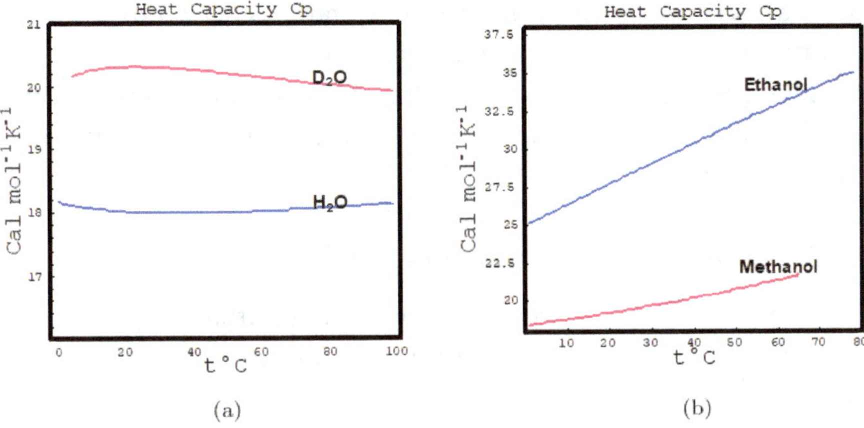

Fig. 1.12.   Heat capacity C_P of (a) H_2O, D_2O, and (b) ethanol and methanol as a function of temperature T at 1 atm.

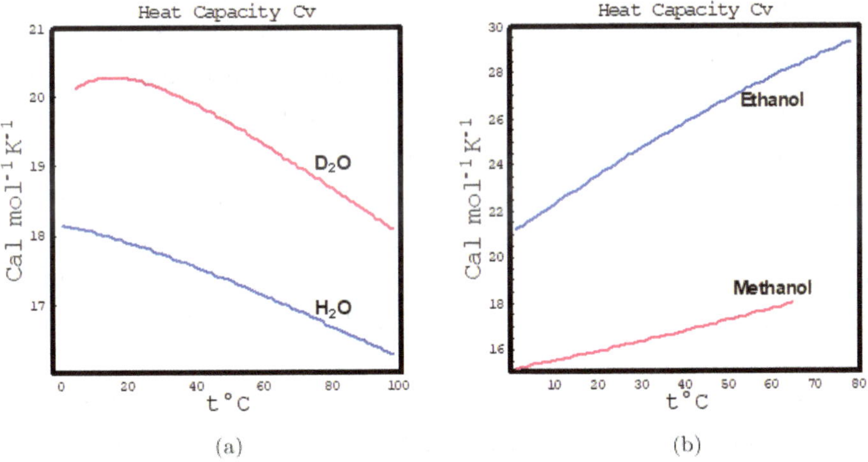

Fig. 1.13.   Heat capacity C_V of (a) H_2O, D_2O, and (b) ethanol and methanol as a function of temperature T at 1 atm.

Figure 1.13 shows the heat capacity C_V for the same four liquids. Note that the curve of C_P of H_2O goes through a minimum at about 35°C. For D_2O there seems to be a maximum of both C_V and C_P around 20°C. Note that the high value of the heat capacity is not unique to liquid water. Other liquids such as ammonia have also high heat capacity.

When we increase the pressure the heat capacities become smaller and the minimum disappears; again, a behavior of a normal liquid.

1.5.3. *Isothermal Compressibility*

The isothermal compressibility of water, denoted as κ_T is a measure to the response of the volume of a liquid to increasing the pressure.[3] This is defined by

$$\kappa_T = -\frac{1}{V}\left(\frac{\partial V}{\partial P}\right)_T = \left(\frac{\partial \ln \rho}{\partial P}\right)_T. \qquad (1.9)$$

Normally, as the temperature of a liquid increases the average intermolecular distance between the particles of the liquid increases. This makes it easier to compress a liquid at a higher temperature. Thus, we expect the compressibility to increase as we increase the temperature (see Fig. 1.14).

Water is anomalous in that it has a region below approximately 45°C, where κ_T actually decreases as the temperature increases. At 45°C, the compressibility passes through a minimum, and thereafter it increases with T, as in the case in normal liquids.

As we increase the pressure, the minimum in the compressibility becomes less and less pronounced, but even at 3000 atm we still observe a shallow minimum.

Table 1.5 shows the isothermal compressibility of some liquids at $P = 1$ atm. Note that water has a relatively small value of compressibility. However, the isothermal compressibility of glycerol at 28° is even smaller at 21.1×10^{-6} atm^{-1}.

---

[3]Some refer to the isothermal compressibility as a "response function". This is true, but it is true for any derivative with respect to P, T or N.

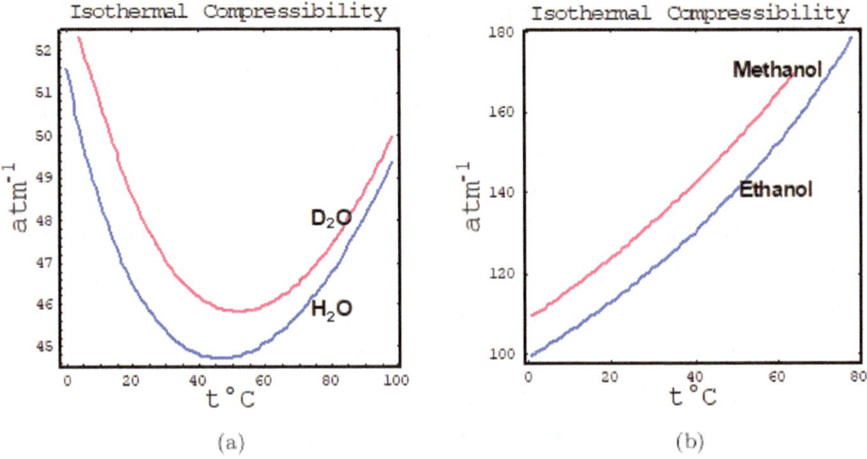

Fig. 1.14.   The isothermal compressibility as a function of t for (a) H$_2$O, D$_2$O, and (b) ethanol and methanol as a function of temperature at 1 atm.

Table 1.5.   Isothermal Compressibility of Water and Some Alcohols.

	$t°C$	$\kappa_T \times 10^6$ atm^{-1}
H$_2$O	20	45.91
	25	45.52
Methanol	20	121.4
Ethanol	20	111.9
Acetone	20	126.2
Ethyl acetate	20	113.2
Ethylene glycol	20	36.4
Glycerol	28	21.1

Liquid water has many other properties that exhibit anomalous values, e.g., the dielectric constant see Appendix D, surface tension, diffusion coefficient, viscosity, etc. Aqueous solutions also exhibit unusual properties, some of which will be discussed in Chapters 3 and 4. In this introduction, we have mentioned only a few properties — the interpretation of which can be given by some simple theoretical arguments.

1.5.4. *The Radial Distribution Function of Water*

The most important experimental information on the mode of packing of water molecules in the liquid state is contained in the radial distribution function, which is obtained from X-ray or neutron scattering data. A solid is characterized by a well-defined structure; the molecules are packed in some orderly fashion. The liquid phase, on the other hand, seems to be lacking any structure. Nevertheless, it is possible to detect some regularities in the packing mode of the molecules in the liquid phase.

If we could sit at the center of one of the particles and observe our surroundings, we would not see any regular pattern that we could call a "structure". But if we count the number of particle centers that appear within a spherical shell of width dR at a distance R, we shall find the following regularities.

Each spherical shell of width dR at a distance R has a volume $4\pi R^2 dR$. If the average (or macroscopic) density of the liquid is $\rho = N/V$, then we should expect to find on average $\rho 4\pi R^2 dR$ particle centers within this volume. However, because our vantage point is located at the center of a specific particle, we observe an average density that deviates from the expected density ρ. We define a function $g(R)$ in such a way that the actual average density at a distance R from the center of a specific particle is $\rho g(R)$. Equivalently, the average number of particles in each spherical shell is $\rho g(R)4\pi R^2 dR$, not $\rho 4\pi R^2 dR$. Thus, $g(R)$ is a measure of the deviation of the actual count of the average number of particles around a given particle, relative to the average number of particles that we should expect to find in a spherical shell of radius R centered at a *random* point in the liquid. The function $g(R)$ is referred to as either the pair correlation function or as the radial distribution function (*RDF*).

Having this qualitative definition of $g(R)$, we now turn to describe some of its salient features for spherical particles.

(i) If σ is the (effective) diameter of each particle in the system, then the probability of finding any other particle at a distance of less than σ from a given particle is almost zero, i.e.,

$$g(R) \approx 0 \quad \text{at } R \leq \sigma. \tag{1.10}$$

The reason for this is that at distances $R \leq \sigma$, the two particles exert strong repulsive forces, hence they are effectively impenetrable. (In fact, the diameter σ is *defined* as the distance, below which the repulsive forces are so large that the two particles practically cannot approach each other to a distance shorter than σ.)

(ii) At a very long distance $R \to \infty$, we expect that the average number of particles in the element of volume dV will not depend on the fact that we have placed a particle at the origin of our coordinate system. Therefore, we expect the average number of particles in the spherical shell around the center of a specific particle to be the same as $\rho 4\pi R^2 dR$. We say that as R becomes very large, there exists no *correlation* between the particles. Clearly, in this case $g(R) \approx 1$, for $R \to \infty$. We have written $R \to \infty$ to signify that R is a very large distance. In practice, for regular liquids not close to their critical point, one finds that the correlation function is very close to 1 (i.e., that there is no correlation) when R is approximately five or six times the molecular diameter of the particles (see Fig. 1.15).

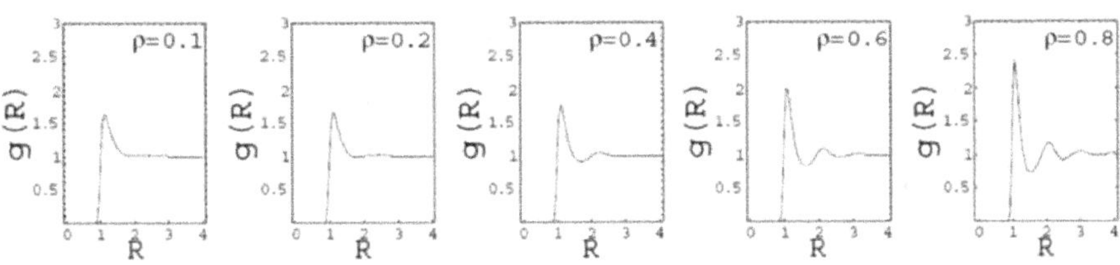

Fig. 1.15. The dependence of the radial distribution function on the density.

(iii) At distances $\sigma \leq R \leq 4\sigma$, we find (both experimentally and from theoretical calculations) that $g(R)$ has, in general, successive maxima and minima, as shown in Fig. 1.15. Note that the first maximum occurs at about $R \approx \sigma$, the second maximum at $R \approx 2\sigma$, and the third at $R \approx 3\sigma$. Each time we increase R, the value of the maximum decreases from $g(R \approx \sigma)$, to $g(R \approx 2\sigma)$, and then to $g(R \approx 3\sigma)$. For $R \approx 4\sigma$ or $R \approx 5\sigma$, the value of $g(R)$ becomes practically unity, i.e., there is almost no correlation beyond say, $R \approx 5$.

(iv) In Fig. 1.15, we see that more maxima and minima appear as the density of the liquid increases. At very low densities, there is only one maximum (left hand curve in Fig. 1.15). This is the limit of $g(R)$ as $\rho \to 0$. At this limit, $g(R)$ is determined only by the pair potential. The relation between the two is: $g(R) = \exp[-\beta U(R)]$. Note that $\rho \to 0$ is the low density limit of a *real* liquid. A *theoretical* ideal gas is a system containing only non-interacting particles. For such particles, $g(R) \approx 1$, i.e., whatever the distances and densities in the system, correlations between the particles are never present.

The typical concentric and approximately equidistant peaks observed in $g(R)$ are a result of the spherical-symmetrical interaction between the particles. This information provided by $g(R)$ is sometimes called the "structure" of the liquid, but when applied to liquid water $g(R)$ is not a good measure of structure.

Figure 1.16(b) shows the first two distances σ and 2σ, where one can expect to find a peak in the pair correlation function for spherical molecules. For non-spherical molecules such as water, the pair correlation function is a function of both the distance and the relative orientation of the pair of molecules. Figure 1.16(a) shows the distances at which we can expect relatively higher densities around a central water molecule.

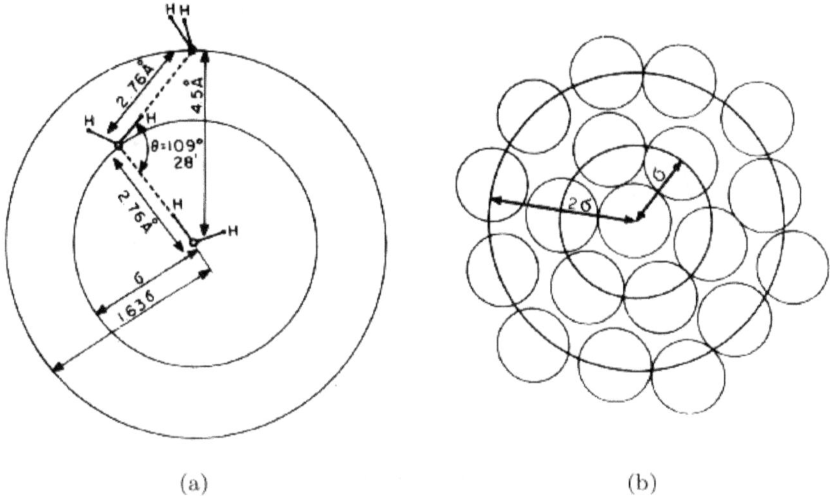

Fig. 1.16. The distribution of first and second neighbors in (a) water, and (b) in simple liquid.

Water, as a hetero-atomic liquid produces a diffraction pattern that reflects the combined effects of O–O, O–H and H–H correlations. Thus, in principle, we have three distinct atom–atom pair correlation functions: $g_{OO}(R), g_{OH}(R)$ and $g_{HH}(R)$. Earlier experimental data based on X-ray scattering could not be resolved to obtain these three functions separately. Therefore, the only information obtained is a weighted average of these three functions. In all our future references to the experimental radial distribution function, we shall always refer to it as $g_{OO}(R)$, or simply $g(R)$. Recently, neutron scattering experiments provided more detailed information on $g_{OO}(R)$, $g_{OH}(R)$ and $g_{HH}(R)$.

Figure 1.17 shows the $g(R)$ for water and argon drawn, not as a function of R, but as a function of the *reduced* distance $R^* = R/\sigma$. The reason we draw the RDF as a function of R/σ is to facilitate the comparison between the two functions. As we have seen, the first peak of $g(R)$ appears at the distance $R \approx \sigma$.

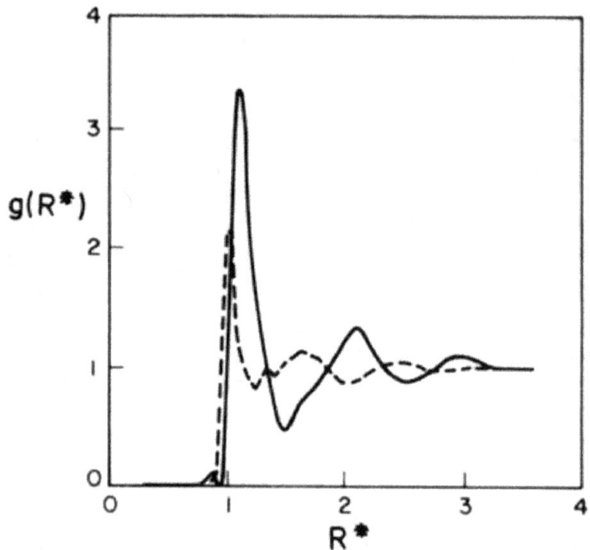

Fig. 1.17.   The radial distribution function for water (dashed line) and argon (solid line) as a function of the reduced distance R^*.

Therefore, plotting $g(R^*)$ brings the first peaks of *RDF* of both argon and water at $R^* \approx 1$.

First, we note that the curves for both argon and water are practically zero for $R^* \leq 1$. Second, the first peaks of the two curves are at about $R^* \approx 1$, and the two curves tend to unity for $R^* \geq 4$, i.e., beyond 4σ there is no more correlation.

The two curves in Fig. 1.17 differ in several aspects that are important to the understanding of liquid water. First, the curve of the water-*RDF* at about $R^* \approx 1$ is narrower, and its maximum value, lower than that of the argon-*RDF*. This is quite surprising. In general, one would have expected the first peak to be higher when the interactions are stronger. This is certainly true if we are in the gaseous phase in which case the *RDF* is given by $g(R) \approx \exp[-\beta U(R)]$. Hence, the stronger the pair interaction the higher the peak. However, in liquid water there is a unique mode of packing the water molecules in such a way that a strong binding energy is correlated with low local density (here binding is simply the total interaction energy of a molecule with its surrounding molecules).

Second, the area under the curve around the first peak is related to the differences in the local density, or the coordination number in the two liquids.

The coordination number is defined using the following procedure. We first choose a radius R_M, then we determine the average number of particles contained within a sphere of radius R_M centered at any given particle in the system. Since $\rho g(R)4\pi R^2 dR$ is the average number of particles at a spherical shell of width dR and radius R, the average number of particles in the *entire sphere* of radius R_M is simply the sum of the average number of particles in all the spherical shells between $R = 0$ to $R = R_M$. This sum is the integral

$$CN(R_M) = \rho \int_0^{R_M} g(R)4\pi R^2 dR. \qquad (1.11)$$

Clearly, once we have $g(R)$, we can calculate $CN(R_M)$. In particular, we can define the first coordination number as the quantity $CN(R_M)$, for a choice of R_M, as the location of the first minimum of $g(R)$, following the first maximum. We shall denote the first coordination number as n_{CN}. The first coordination number for water at 4°C and 1 atm was estimated to be around 4.4. For liquid argon at 84.25°K and 0.7 bar, the first coordination number is about 10. The latter value is typical of a closed-packed liquid having a coordination number in the range of 10 to 12. The small value of 4.4 for water is unusual for a liquid, and is strongly reminiscent of the exact coordination number of the solid hexagonal ice I_h. It suggests that the "local structure" of water is very similar to the local structure of ice. No similar resemblance in local structure is observed in argon, or in any other simple liquid. It is also of interest to note that the first coordination number of a simple liquid normally decreases with the increase of temperature. The opposite is true for liquid water.

The third difference between the two *RDF* curves in Fig. 1.17 is the location of the second peak in the *RDF*. We have noted that a large correlation can be expected at about $R^* \approx 2$ (i.e., at a distance of $R = 2\sigma$) for a simple liquid such as argon. Indeed, we see that the *RDF* of argon has a peak at about $R^* \approx 2$. This is "normal" behavior, as we have explained before. In contrast, the *RDF* of water shows a second peak at an "abnormal" location at about $R^* \approx 1.6$, i.e., at a distance of $R = 1.6\sigma$. If we take $\sigma \approx 2.8\,\text{Å}$ as the effective diameter of water, we have the second peak at $R = 4.5\,\text{Å}$. What does this distance mean in terms of the packing mode of water molecules?

We recall that in the structure of ordinary ice (see Sec. 1.3) the O–O distance is $2.76\,\text{Å}$. A simple calculation shows that the second-nearest-neighbor distance between oxygen atoms is $2 \cdot 2.76\sin(\theta_T/2) = 4.5\,\text{Å}$. This is the ideal O–O distance for second-nearest-neighbor oxygen atoms in ice. The fact that the second peak in the *RDF* of water is also found at $4.5\,\text{Å}$ means that the probability of finding second-nearest-neighbors in liquid water is greatest at $4.5\,\text{Å}$. This, together with the coordination number of water is a strong indication that the local structure around a water molecule is very similar to the local structure around a water molecule in ice I_h.

These three features of the radial distribution function lead to the following conclusion: The basic geometry around a single molecule in water is, to a large extent, similar to that of ice. This is to say that on the average, each molecule has a coordination number of about four, and furthermore, there is a high probability that triplets of molecules will be found with nearly the same geometry as triplets of molecules at successive lattice points in ice. This conclusion pertains only to the *local* environment of a water molecule; no information whatsoever is furnished by $g(R)$ on the structure of the extended layer of molecules. In other words, if one were to sit at the center of a water molecule and observe the local geometry in a sphere of radius, say 5 Å, one would most of the time see a picture very similar to the one seen from an ice molecule, with frequent distortions caused by thermal agitation typical of the liquid state. These facts were recognized long ago by Bernal and Fowler, and by Pauling. Beyond that radius (of about 5 Å), the structure will not be recognized as the one of ice.

1.6. The Importance of Water in Biological Systems

Water affects in specific ways a multitude of biochemical processes taking place in each of our cells. It has long been speculated that life as we know it on this planet could not have started without water. Although this is merely a speculation, it is a reasonable one. Strictly speaking, of course, we cannot preclude the possibility of the existence of some other kind of life form based on some other liquid.

1.6.1. *The Macro-Properties of Water and Life*

As we have noted, the molar volume of ice is larger than that of liquid water. Everyone is familiar with the phenomenon of ice floating on top of water. If the molar volume of ice were *not* larger than that of water, ice formed on the surface of the ocean would sink to the bottom and then be thermally insulated by the water covering it (water is a good thermal insulator). Over time, more and more layers of ice would accumulate at the bottom in a process that would gradually freeze nearly the entire ocean.

The fact that ice floats on top of water helps to maintain a relatively high temperature of liquid water underneath. This in turn helps to maintain life in the water in spite of the lower temperature above the surface. Clearly, when the temperature of the atmosphere increases, it is the ice on the surface that melts first. This also contributes to the maintenance of the water temperature underneath.

The fact that ice has a larger volume than liquid water can also be harmful to living organisms. When a living cell containing water freezes, the expanding ice can break the cell's membrane, possibly killing the cell. Some marine creatures that live in extremely cold environments possess a specific "antifreeze" protein that lowers the freezing temperature and delays the process of ice formation.

The high value of the heat capacity of water is of primary importance in regulating the temperature of a living system. In each living cell, thousands of chemical reactions take place. Some of these chemical reactions are exothermal, i.e., they release heat into the cell's surroundings. If this heat is not absorbed by some mechanism, the temperature will rise to a dangerous level. The large heat capacity of water means that for a given amount of heat absorbed by one unit volume of water, the temperature increase is smaller than it would be in a normal liquid. Thus, most of the heat released in chemical reactions is absorbed by the water — with a minimal increase in its temperature. This is also the reason why laboratories use water as a thermostat-bath when a constant temperature must be maintained. Because of its large heat capacity, the temperature does not change so rapidly as a result of heat flowing inside or outside of the thermostat.

Another property of water that helps regulate body temperature is its large heat of vaporization. When the atmospheric temperature is high, both animals and plants evaporate some of their water content. This is the familiar phenomenon of sweating. Because the heat of the vaporization of water is large, only a small quantity of water needs to be evaporated to maintain the temperature of the body. Note that the actual amount of sweat visible on the skin in hot weather or after exercising is not an indication of the amount of water that is evaporating. The evaporating sweat is released into the air as a gas and is not conspicuous. Sweat becomes visible only when liquid is released by the body more rapidly than the rate of evaporation. The rate of evaporation is affected by the relative humidity of the air. When the relative humidity is low, water evaporates quickly from the skin. When it is high, water evaporates slowly from the skin because the air has a lowered capacity to absorb additional water, and temperature regulation by sweating becomes inefficient. This is the reason one feels more comfortable in the dry atmosphere of Jerusalem than in the humid atmosphere of Tel Aviv.

The large heat of vaporization helps to regulate not only the bodies of living organisms, but also the Earth's natural environment. Most of the radiant energy from the Sun that reaches the Earth is absorbed by evaporating ocean water. The fact that the heat of evaporation of water is large partially explains why this evaporation effectively prevents an increase in the Earth's temperature. If the seas were filled with some other liquid, e.g., alcohol, the Earth's temperature would sharply increase. Such an increase would probably kill all living creatures (assuming they had developed in, and were surviving in an "alcoholic" environment, which is a doubtful possibility).

We have mentioned the large amount of heat that is absorbed when ice melts, and the damaging effect of any water freezing within a living cell. If the temperature of pure water or any aqueous solution falls below the freezing point (which is normally lower for aqueous solutions than for pure water), ice begins to form and a large amount of heat is released. This release of heat decelerates the rate of any further ice formation, preventing, or at least delaying the conversion of all the water in the body into ice.

In concluding this section, we mention two more properties of water that are important in biological systems. The first is a critical factor affecting the ability of plants to absorb water. This property is surface tension. Compared to "normal" liquids (e.g., ethanol, hexane and benzene), water has a very high surface tension. The surface tension of water at 20°C is about 72.75 (in ergs cm^{-2}), compared with acetone (23.7), ethanol (22.3) and n-hexane (18.4) at the same temperature. This is significant because surface tension is the primary force that draws fluids up through capillaries. The higher the surface tension of a liquid, the higher the liquid can rise in capillaries. This explains why water can rise up in the interior of tall trees, even up to the highest leaves where it is needed in photosynthesis and temperature regulation.

Another important property of water is the dielectric constant (see Appendix D). The dielectric constant Appendic D of water exceeds that of many liquids, for instance, hexane (1.87), chloroform (5.05), ethanol (24), methanol (33), and water (80). However, hydrogen cyanide (116) exceeds the dielectric constant of water. In order to dissolve a solute, the solute–solute interactions must be broken. The larger the dielectric constant of solvent, the weaker such interactions are (recall that the force between two charges is: $F = q_1 \cdot q_2 / \varepsilon \cdot r^2$ where q_1 and q_2 are the charges of the interacting solutes, r is the distance between the two charges and ε is the dielectric constant). Solvents with large dielectric constants would form stronger ion–solvent interactions and be more likely to dissolve ionic compounds (see Appendix B). A more extensive discussion of the relevance of the properties of water to life may be found in (Henderson, 1913; Edsall and Wyman, 1958; Franks, 2000).

1.6.2. *The Micro-Properties of Water and Life*

With our macroscopic eyes, liquid water looks like a simple structureless medium in which life activity takes place. The situation is very different when we look at water with our "microscopic eyes", i.e., when we enter the microscopic world at the molecular level.

Water molecules form a kind of *interacting society*, resembling that of busy ants swarming around a mound of sand or bees in the hive. Water is by no means a continuum as it looks to us in the macroscopic world. Let us start a journey inside water.

We can compare the *social behavior* of water molecules to that of individual ants in an ant colony. They perform a particular task only when a corresponding stimulus from their environment exceeds a certain value. Water molecules start interacting with biological molecules when they approach the surface of these molecules. Then, water molecules movement is no longer chaotic and a new molecule structure is formed; the solvated biological molecule which is ready to accomplish specific biological functions.

At the macroscopic level, water in a glass on a table remains at rest while water in rivers flow at different speeds, depending on whether we are considering either a river flowing through the mountains close to its source, or a river approaching its mouth. However, at the microscopic level, water molecules are moving all the time — vibrating, rotating and changing their locations.

Depending on their location, water molecules may be organized, show a variety of structures, or may even be bound to a specific location of a biological macromolecule.

As will be shown through the pages of this book, the motion of water molecules is quite different in bulk water or when other substances are present. This latter situation occurs when we consider solutions or when water is forming part of the cells or the blood flow. The structure and dynamics of water is remarkably different depending on the environment.

In the following chapters, we will enter the incredible and fascinating world of water, which is full of yet unresolved puzzles intimately related to the biggest mystery of the universe — the enigma of life, the greatest challenge to science.

Let us first start by taking a look at the unseen activity occurring in the entirely still water in a glass. Did we say unseen?! Yes, as far as we deal with pure water. But let us add some tiny (colloidal) particle (3 to 1000 nm in diameter) into the glass of water. A colloidal particle is much larger than solvent molecules, but small enough as to remain suspended despite gravity force. It can be observed only through a microscope, moving randomly, aimlessly, and constantly changing directions (see Fig. 1.18). What drives this hectic "crazy" movement? What is going on within the calm water? This was the first experimental indication of energetic activity within water, which we now know to be consistent with the idea of molecules moving around, hitting the tiny particle. The colloidal particle is bombarded randomly by the molecules of water. This is the so-called Brownian motion, discovered by the botanist Robert Brown in 1827. The mechanism responsible for this phenomenon was first suggested by Albert Einstein in 1905, who explained in detail by means of a mathematical model how the motion of the microscopic particle provoked random collisions with individual water molecules. Einstein derived the following equation:

$$\langle x^2 \rangle = \frac{k_B T}{3\pi\eta}t = \frac{RT}{3\pi\eta r N_A}t,$$

where k_B is the Boltzmann constant, R is the gas constant, T is the temperature, η is the viscosity of the solvent, r is the radius of the colloidal particle, N_A is the Avogadro number, and $\langle x^2 \rangle$ is the mean square displacement of the colloidal particle.

Until then, the existence of atoms and molecules as the basic constituents of matter had only been hypothesized and strongly debated. Great scientists like Wilhelm Ostwald defended the position that atoms and molecules were merely mathematical fiction. According to Einstein's equation on Brownian motion, the measurement of the mean square displacement of the colloidal particle would allow for the estimate of the Avogadro number. Einstein showed that the overall visible motion of the "big" particle, when averaged over many observations, exactly matched what would be expected if the little particles batting the "big" microscopic particle were indeed atoms or molecules. In 1908, Jean Perrin, using a dark-field microscope, experimentally verified Einstein's equation, getting correct values for the Avogadro number, thus providing experimental confirmation of the real existence of atoms and molecules. Ostwald told Arnold Sommerfeld, one of the brightest scientists at that time, that he had been converted to atomistic belief by the complete explanation of Brownian motion. Vigorous controversies like this paves the way for science to make progress in the right direction and under solid ground.

What we see in Fig. 1.18 is an illustration of water molecules hitting a tiny microscopic particle, which is comparatively huge compared to them.

The direction of the force on the microscopic particle due to the bombardments of water molecules changes constantly. These bombardments make the particle move randomly at the macroscopic level.

It should be mentioned that the experimental observations by Brown and the theoretical prediction of Einstein (that were experimentally confirmed by Perrin) were decisive in convincing

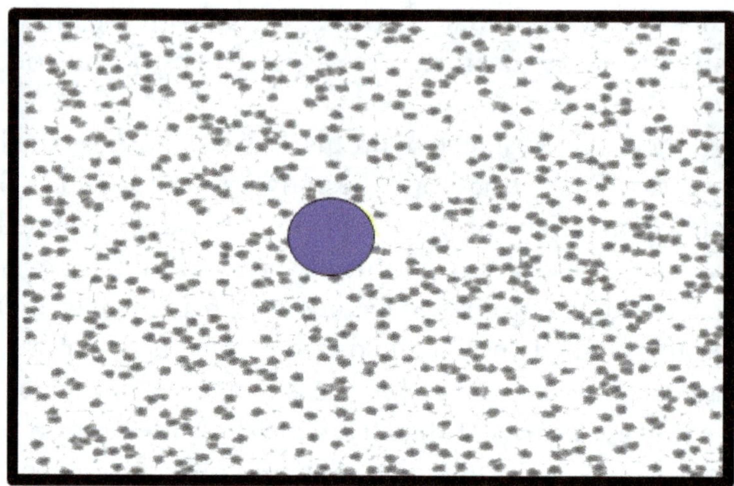

Fig. 1.18. Water molecules surrounding and randomly hitting a microscopic body.

the scientific community of the reality of the atomic nature of matter.

Water molecules are not dots. Their detailed shape and structure will be described later. We need to investigate the interaction between water molecules and other biological molecules existing in the cell. Remember that water molecules are present both inside the living cell and between cells. Water molecules also regulate the activity of biological molecules and biological processes. How do they do it? We shall discuss this question later in this book. We have realized previously that the motion of water molecules can be fast or very slow in different regions. We shall later demonstrate by computer simulation how the motion of water molecules depends on the environmental biological molecules.

The other aspect of water molecules we shall discuss is the details of their structure and interactions. We shall "zoom in" to observe the detailed structure of water molecules, which will not be presented as "dots" anymore.

Now we know that water at the molecular level plays an active and even decisive role in all living cells' activities. Water does not play the passive role of a container of biomolecules but it also plays a dynamical role.

The 2016 Nobel Prize in Chemistry was awarded to Jean-Pierre Sauvage, J. Fraser Stoddart and Bernard L. Feringa for the design and synthesis of molecular machines, which can be controlled to accomplish a given task once they are supplied with energy.

It has been shown that for molecular machines with hydrogen-bonded components, water acts as a kind of lubricant. Matthijs R. Panman and coworkers have recently proposed (*Nature* 5: 929–934, 2013) that the accelerating effect of water is due to its ability to form a 3D hydrogen-bond network between the moving parts of the molecular machine. According to these authors, their results might indicate a more general phenomenon that helps to explain the function of water as the "lubricant of life".

In this manner, water makes the molecular machines work in the microscopic world in a similar way as if it were mainly responsible for the operation of industrial machines in past centuries.

Before we discuss water in the surrounding of macrobiological molecules, let us first take a look at the structure of water molecules surrounding certain relatively small organic molecules — benzene (C_6H_6) and cyclohexane (C_6H_{12}) — called hydrocarbons because their formulas only contain carbon and hydrogen atoms. In the next chapters, we will show how quantum mechanics and statistical mechanics help in getting a "real picture" for these molecular associations. Let us by now just briefly analyze the results of the calculations.

Figure 1.19 shows the structure of the benzene-water molecular complex as resulting from quantum calculations. We see

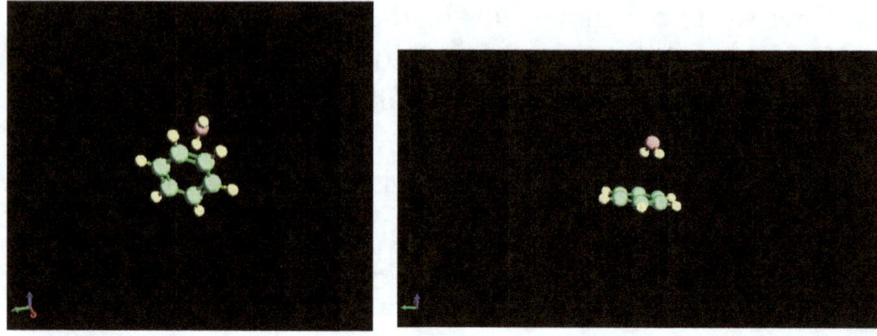

Fig. 1.19. Two views of the benzene-water complex from quantum calculations.

that the water molecule interacts with the aromatic hydrocarbon through what is technically known as a van der Waals interaction — a kind of bonding involving the oxygen electrons and the so-called *π electron cloud* in benzene (quantum calculations show that a number of electrons, called π electrons, in aromatic hydrocarbons, are delocalized around the ring formed by the carbon skeleton). The complex is very weak, with a bonding energy of about 2–3 kcal/mol. Calculations on the corresponding molecular complex between water and cyclohexane fail to locate any stable molecular association. Colloquially, we can say that water "hates" hydrocarbons.

The different kinds of interaction between benzene and cyclohexane with water is manifested in the snapshots from molecular simulations collected in Fig. 1.20. While we observe the presence of a bubble surrounding cyclohexane with no water molecules, in the case of benzene, water molecules in the z-axis close to the center of the ring do penetrate the bubble.

On the contrary, when a long flexible molecule like a protein is surrounded with water molecules there is an affinity of water molecules to some parts of the molecule surface, which guides the molecule to get its final configuration surrounding the protein.

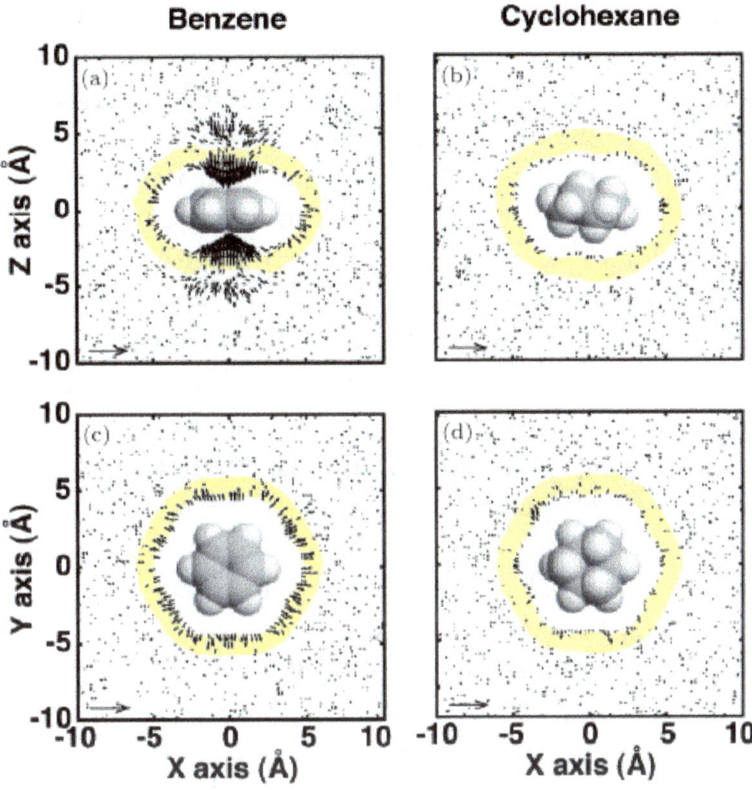

Fig. 1.20.   Simulation of water molecules around organic molecules.

The behavior of water molecules in both cases is dictated by electric attraction or repulsion between molecules, as will be analyzed later.

When water molecules approach a protein, they surround or penetrate it in a very peculiar manner by "dancing" around it and changing its shape, thus enabling its folding to the precise configuration required to play its biological function.

Water molecules surrounding proteins and DNA play a vital role in determining their structure and functions. Water molecules and protein regions both play a concerted dance dictated by their mutual chemical interactions. Details of the structural, dynamical and energetic aspects of this "great show" will be analyzed later in the book.

One should keep in mind that every atom and molecule surrounding the protein is in a continuous motion. Motion of water molecules includes rotation and vibration. Atoms of the protein vibrate too and rotate as far as they can (depending on its bonding constraints). The main factor controlling the magnitude of the motion is temperature.

The role of water in biological science starts when water comes close to a biomolecule's very complicated surface, possibly penetrating it, and in a way, becoming virtually an organic part of the biological molecule, thus affecting its structure, conformity and functionality.

The inner regions in a protein are tightly packed but do include a number of cavities. Water molecules, intruded into those cavities, promote stability in the protein conformation. This is due to "loving" linkages between water molecules and particular regions of the protein chains; the hydrophilic region is the technical term. You will learn more in Chapter 3 about "loving" (hydrophilic) binding, as well as the "hating" (hydrophobic) repulsion. Both are responsible for the structure and functionality of biological molecules.

To summarize, proteins would not have performed their function had they not been surrounded by water molecules forming what is known as the hydration shell.

Water also plays an essential role in the complex structure of the cell by mediating intramolecular interactions in molecules and intermolecular interactions in between molecules, as in protein-DNA interface, where they are abundant.

Chromosomes are complex structures involving DNA and special proteins called histones. The genome of a human cell is located within 46 DNA huge molecules inside the cell nucleus. Its length (stretched) is about two meters. However, the cell nucleus has a diameter of about 10^{-5} m. Therefore, DNA must be compressed accordingly. Histones play the role of packing proteins.

Water molecules also play a major role in determining the mode of packing of DNA.

Proteins protect their folded shape and resist unfolding to preserve their biological functionalities. Random gene mutations might produce alterations in the hydration structure of proteins, thereby lowering its resistance to unfolding. The whole process gives rise to a wrong shape to work properly (loss of functionality). This may be avoided by clinging proteins to each other, thus protecting water sensitive areas of contact against surrounding water.

Here we meet a key feature of cell biology: Association between proteins. It has important functions in sending molecular signals through a cell membrane. The way proteins interact with each other may have been selected by evolution. Water may then have a role in this evolutionary process.

1.6.3. *Conclusion*

The fact that the properties of water are "fine-tuned" to support life is sometimes viewed as evidence of the existence of an "intelligent designer". This is essentially the same argument as the so-called anthropic principle invoked by creationists in favor of the existence of an "intelligent designer": If the universe is so finely tuned to support life, there must be a "fine-tuner" who designed the universe. In other words, an intelligent being has *designed* the universe in such a way so as to support life.

This type of argument is quite appealing to those who believe that life, as we know it on our planet, is the only possible form of life. However, one cannot exclude the possibility that other forms of life based on other liquids, or even on different "fine-tuned" laws of physics, could have evolved in some remote part of the universe, or perhaps in other universes.

In Henderson's *The Fitness of the Environment* (1913), the author discusses the reciprocity of Darwin's fitness ideas, and the fitness of water to life. He asks the question: "Water is indeed a wonderful substance which fills its place in nature most satisfactory, but would not another substance do as well? Is not ammonia, for example, a possible substitute?"

Henderson concludes the chapter on water with the following words:

"In truth Darwinian features is a perfectly reciprocal relationship. In the world of modern science, a fit organism inhabits a fit environment."

CHAPTER 2

Water Molecule: Structure and Interaction

Water molecules have so far been depicted in the previous chapter as dots or electrical dipoles at most, ignoring any structural detail, and nothing has been said about the source and character of the interaction between molecules.

2.1. Theoretical Tools to Analyze Molecular Structure

In this chapter, we discuss some of the basic concepts in theoretical chemistry, such as molecular geometries and energies. It is therefore advisable to provide a brief account of the theoretical tools available in order to assimilate that kind of information.

2.1.1. Elemental Notions on Quantum Calculations

To proceed in the simplest way, we will focus on a very elemental physical system: A harmonic oscillator, i.e., a mass on a spring!

In classical mechanics, when displaced from its equilibrium position, the 1D harmonic oscillator experiences a restoring force that according to Hooke's law is proportional to the displacement of the opposite sign:

$$F = -kx = -\frac{\partial V}{\partial x} \quad \text{or} \quad V = \frac{1}{2}kx^2, \tag{2.1}$$

where k is a constant and V is the potential energy.

Newton's second law is:

$$F = ma = m\frac{d^2x}{dt^2}. \tag{2.2}$$

The solution of the above equation is easy. It provides information about the motion of the particle of mass m.

$$x(t) = A\cos 2\pi v(t - t_0), \quad \text{with } v = -\frac{1}{2\pi}\sqrt{\frac{k}{m}}, \tag{2.3}$$

where A is the amplitude (maximum elongation), v is the frequency (number of cycles per unit time) and t_0 is a constant that allows the origin of time to shift. Equation (2.3) provides the dependence of the location of the particle as a function of time.

It is easy to show that the total energy E and the velocity $v(t)$ of the system are given by:

$$E = 2m\pi^2 v^2 A^2 \quad \text{and} \quad [v(t)]^2 = 4\pi^2 v^2 A^2 \sin^2 2\pi v(t - t_0). \tag{2.4}$$

The mass m moves in a periodic way from $x = 0$ (minimum displacement) to $x = A$ (maximum displacement).

It is important to stress that: (a) A classical oscillator can have any energy depending upon the amplitude of vibration, and it can be measured, and (b) there exist regions that are forbidden. Indeed, Fig. 2.1 shows that the points outside the potential curve, like P_1, are not classically allowed because $V_1 > E_1$, i.e., the potential energy is greater than the total energy. This would imply that the kinetic energy K should be negative, thus violating classical physics.

Let us now briefly summarize the main results of a quantum mechanical treatment of the same problem. The starting point is Louis de Broglie's revolutionary idea (at that time) that every particle of mass m and velocity v carrying a fixed momentum $p = mv$ is accompanied by a wave of a length λ

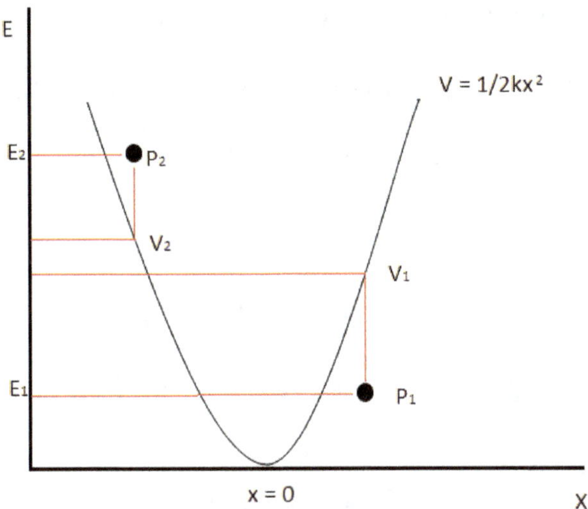

Fig. 2.1.   Allowed (P_2) and forbidden (P_1) regions for the case of the classical 1D oscillator. E and V stand for total and potential energies, respectively.

equal to $\frac{h}{p}$ (where h is Planck's constant), which is approximately 6.63×10^{-23} joule-second. De Broglie had been inspired by the new corpuscular theory at his time where monochromatic electromagnetic radiation appeared to be "composed" of "particles" characterized by having an energy and momentum given by:

$$E = h v, \quad p = h/\lambda$$

Defining $\omega = 2\pi v$ and $k = 2\pi/\lambda$. We find:

$$E = \hbar\omega \quad \text{and} \quad p = \hbar k. \tag{2.5}$$

According to de Broglie, the electron is described by a wave and a whole number of wavelengths must fit along the circumference of the electron's orbit $n\,\lambda = 2\pi r$ which explains the discrete electronic orbits of electrons in the atom. A free particle with defined momentum is then represented by:

$$\psi(x,t) = A[\cos(kx - \omega t) + i\sin(kx - \omega t),$$

$$i = \sqrt{-1}, \quad k = \frac{2\pi}{\lambda}. \tag{2.5-1}$$

In short notation:

$$\Psi(x,t) = A\exp[i(kx - \omega t)]. \tag{2.6}$$

This equation describes a free particle of well-defined k and energy

$$E = \hbar^2 k^2 / 2m. \tag{2.7}$$

Equation (2.6) can be written in the form:

$$\Psi(x,t) = e^{-\frac{iEt}{\hbar}}\psi(x), \tag{2.8}$$

$$\psi(x) = e^{ikx}. \tag{2.9}$$

So far, so good, mathematically. But what is the meaning of all this? What does ψ mean? The complex conjugate of the function $\psi(x)$ is obtained by replacing i by $-i$ and is denoted by ψ^*. Then:

$$\psi^*(x)\psi(x) = |\psi|^2. \tag{2.10}$$

Here comes a main postulate of quantum mechanics that relates the so-called amplitude of matter field to experiment:

The probability of finding the particle described by the wave function $\psi(x)$ in the interval dx around the point x is $|\psi(x)|^2 dx$.

In other words, the probability *per unit length* (or *probability density*) of finding the particle at x is $|\psi(x)|^2$.

For plane wave (2.9), $|\psi(x)|^2 = 1$ means that the probability of finding the particle is the same at any point. However, if we combine wave functions of different wave numbers like that of:

$$\psi(x) = \int A(k)e^{ikx}dk, \tag{2.11}$$

instead of single wave (2.9), where $A(k)$ is the weight given to each k determining the range of k values Δk entering the summation and the range of x value in which $|\psi(x)|^2$ has considerable value, is also limited to forming a "wave packet" concentrated mainly in the interval Δx

We shall come back later to the relation between Δx and Δk.

But how is a wave function created in a real situation like a particle under the influence of a potential field $V(x)$ in a molecule and involved in chemical bound? What replaces the classical Newton equation of motion? Coming back to the *time-dependent* wave packet:

$$\Psi(x,t) = \int A(k)\exp i(kx - \omega t)dk. \tag{2.12}$$

Then:

$$\frac{\partial \Psi(x,t)}{\partial t} = -i \int \omega A(k)\exp i(kx - \omega t)dk, \tag{2.13}$$

$$\frac{\partial^2 \Psi(x,t)}{\partial x^2} = - \int k^2 A(k)\exp i(kx - \omega t)dk. \tag{2.14}$$

For a free particle of well-defined energy $E = \hbar\omega = \hbar^2 k^2/2m$.

We conclude that:

$$i\hbar\frac{\partial \Psi(x,t)}{\partial t} = \frac{\hbar^2}{2m}\frac{\partial^2 \Psi(x,t)}{\partial x^2}. \tag{2.15}$$

This is the time-dependent Schrödinger equation for a free particle.

Considering a particle moving in a field of force corresponding to potential energy $V(x)$, meaning that the total energy is

$\frac{p^2}{2m} + V(r)$, this suggests that the Schrödinger equation for a simple particle of mass m moving in one dimension should be:

$$-\frac{\hbar}{i}\frac{\partial \Psi(x,t)}{\partial t} = \frac{\hbar^2}{2m}\frac{\partial^2 \Psi(x,t)}{\partial x^2} + V(x) \cdot \Psi(x,t), \qquad (2.16)$$

where $\Psi(x,t)$ is the wave function that as quantum mechanics postulates, contains all the information we need to know about the system it describes.

Expression (2.16) is the master equation in quantum mechanics, as is Newton's equation (2.2) in classic mechanics.

A general solution to Eq. (2.16) for a potential function $V(x)$, *which is not* a function of time, has the form:

$$\Psi(x,t) = e^{\frac{-iEt}{\hbar}}\Psi(x). \qquad (2.17)$$

The state of a system described by $\Psi(x,t)$ in Eq. (2.17) is said to be a stationary state, and E represents the energy of the system that can be shown to be a constant. (We are considering here what is known as conservative systems; those where the force is derivable from a potential energy function. See Eq. (2.1).)

By substituting Eq. (2.17) into the time-dependent Schrödinger equation (2.16), one gets the time-independent Schrödinger equation:

$$-\frac{\hbar^2}{2m}\frac{d^2 \Psi(x)}{dx^2} + V(x)\Psi(x) = E\Psi(x), \qquad (2.18)$$

whose solution provides both the wave function $\Psi(x)$ and the energy E of the system considered.

It is not necessary to provide further details here on the particular solution of Eq. (2.18) for the case of the 1D harmonic oscillator. The interested reader can consult any quantum mechanics textbook. After some algebraic manipulations one finds:

$$\Psi_n(x) = N e^{\frac{s^2}{2}} (-1)^n \frac{d^n}{dx^n} e^{-s^2}, \quad n = 0, 1, 2, \ldots \tag{2.19}$$

$$E_n = \left(n + \frac{1}{2}\right) h\nu, \tag{2.20}$$

where N is a normalization constant and $S = \alpha x$, with α being a constant (see Fig. 2.2).

The wave function $\Psi_n(x)$ contains the information about the system, specifically, $|\Psi_n(x)|^2$ provides information to find the particle at location between x and $x + dx$. In the case of the classical harmonic oscillator, we have obtained equations that provide the energy of the system (see Fig. 2.2). Let us

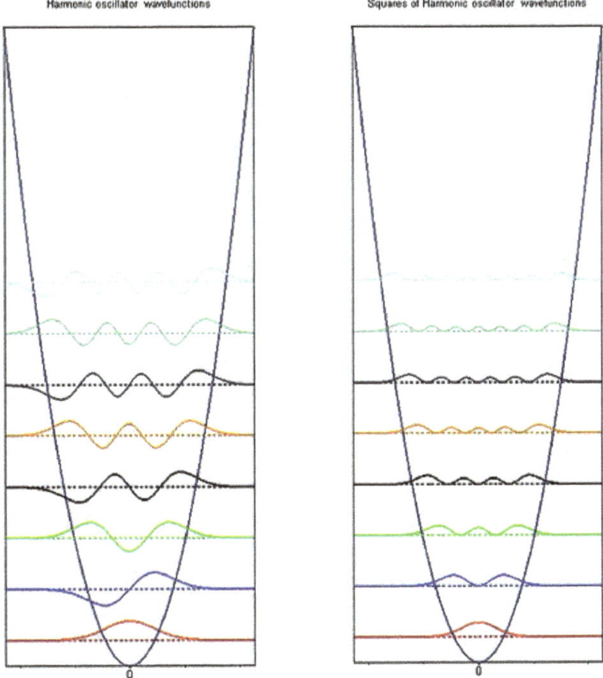

Fig. 2.2. Probability densities $|\Psi_n(x)|^2$ and energies, solution of Eq. (2.18). See Eqs. (2.19) and (2.20).

now compare both solutions to learn about some fundamental differences between the classical and the quantum mechanical solutions.

The first important difference between classical and quantum solutions is that energy is quantized in the latter, i.e., quantum energy cannot adopt any value as in the classical oscillator (see Eq. (2.4)) where all positive values are allowed. In the case of the quantum oscillator, the allowed energy values are $1/2, 3/2, 5/2, \ldots$ times $h\nu$ (see Eq. (2.20)). While energy solutions of the Newton equation are continuous, Schrödinger equation energy solutions are quantized (n in Eqs. (2.19) and (2.20) is an integer number). The new mechanics, developed through the first third of the 20th century, adopted its current name — quantum mechanics — to highlight that paradigmatic new finding.

Another difference is that the quantum energy for the ground state of the harmonic oscillator (that of the lowest energy, namely, $(1/2)h\nu$, see Eqs. (2.19) and (2.20) and Fig. 2.2) is not zero, in sharp contrast with the classical treatment. This striking result, the zero-point energy, is a direct consequence of Werner Heisenberg's uncertainty principle, which states that the product of uncertainties in the momentum (Δp) and the position (Δx) of a particle fulfil the inequality

$$\Delta p \Delta x \geq \hbar/2. \qquad (2.21)$$

Therefore, in the new mechanics, we cannot simultaneously estimate exact values for the momentum and the position for a particle. The more precise the momentum is, the larger the error in the position. It is important to stress that the uncertainty principle is not a limitation inherent in the measuring tools available for the experiments, which will be minimized with the continuous technological progress. It is a law of Nature that forces us to abandon the classical deterministic world for a new probabilistic scenario.

Albert Einstein refused to accept the uncertainty principle and worked tirelessly for finding arguments against it. Niels Bohr refuted all of them, one after another. The prominent physicist Richard Feynman emphasized that the complete theory of quantum mechanics depends on the correctness of Heisenberg's uncertainty principle. In this sense, it can be considered as the cornerstone of physics.

A third important difference refers to the fact that it can be shown that, according to the quantum solution, there is some probability of finding the harmonic oscillator mass at any point on the x-axis (except the nodes; see Fig. 2.2). As mentioned previously (see Fig. 2.1), there are classically forbidden regions, namely those where $V > E$ (see Fig. 2.1), as it would imply negative kinetic energies. The presence of quantum particles in this classically forbidden region is called quantum tunneling.

Finally, it is important to remark that the so-called probability densities, given by:

$$\Psi^*(\mathbf{r}, t)\Psi(\mathbf{r}, t) = |\Psi|^2, \tag{2.22}$$

where $|\Psi(\mathbf{r}, t)|^2 d\tau =$ the probability of finding a particle in volume element $d\tau$ at point $\mathbf{r}$ and at time t, for quantum stationary states are seen to be markedly different from those expected classically. According to classical mechanics, the particle is most likely to be found at the extremities of its motion where its velocity is the lowest. Figure 2.2 shows that this is not true for the lower states of the quantum harmonic oscillator. However, as n increases, the quantum mechanical picture rapidly begins to approach one that is obtained from classical mechanics (see $n \to \infty$ in Fig. 2.2). This is a particular case of the so-called correspondence principle formulated by Bohr in 1920 — the behavior of systems described by quantum mechanics reproduces classical results in the limit of large quantum numbers.

Quantization, zero-point energy, uncertainty principle and tunneling are phenomena unequivocally confirmed by experience and do form part of many of the technological advances we enjoy at present. They are also the basis for future advances currently under development in many laboratories all around the world. Feynman summarized the situation at the end of last century: "*Physics has given up. We do not know how to predict what would happen in a given circumstance, and we believe now that it is impossible that the only thing that can be predicted is the probability of different events. It must be recognized that this is a retrenchment in our earlier ideal of understanding nature. It may be a backward step, but no one has seen a way to avoid it... We suspect very strongly that it is something that will be with us forever, that it is impossible to be at the puzzle, that this is the way nature really is*".

As a resume of this section, we can conclude that given a microscopic system (atom, molecule, molecular aggregates, etc.), the solution of the Schrödinger equation provides the energy E of different stationary states allowed for the system and the wave function $\Psi(r_1, r_2, \ldots, r_N)$, which is a function of the coordinates of the N electrons in the system. In general, Ψ can be complex (Ψ^* will be the Ψ conjugate). Of course, the larger the number of electrons in the system, the greater the computational difficulty to solve the Schrödinger equation. The continuous evolution of computational technology allows us the consideration and solution of more and more complex chemical and biological systems, as we will have the opportunity to see in the next sections. As mentioned previously, the knowledge of the wave function $\Psi(r_1, r_2, \ldots, r_N)$ for a system allows for the calculation of all its properties. The interested reader can consult any textbook on quantum mechanics to follow all the mathematical details required. As regards, to this book, all that the reader should know is that all properties that will be discussed in the next pages can be computed from $\Psi(r_1, r_2, \ldots, r_N)$, once the corresponding Schrödinger equation is properly solved.

2.1.2. *Elemental Notions on Computer Simulations*

Martin Karplus, Michael Levitt and Arieh Warshel won the 2013 Nobel Prize in Chemistry for the development of multiscale models for complex chemical systems.

In multiscale modeling, multiple models at different scales are employed simultaneously to describe a given system. In order to perform simulations of chemical reactions involving large (complex) biological molecules, the small reactive areas (usually tens of atoms) can be treated at higher (computationally more expensive) theoretical levels (quantum-mechanics (QM)) while the chemically "inert" regions (thousands of atoms) are described at the less sophisticated classical molecular-mechanics (MM) (classical techniques with no wave functions) level (much less computationally demanding), which empirically (or semi-empirically) employs parametrized force fields. The resulting multiscale modeling is called the hybrid QM/MM approach.

The introduction of the QM/MM model to tackle molecular dynamics (MD) simulations in biomolecules was made by Warshel and Levitt in a seminal paper on the theoretical studies of enzymatic reactions, which appeared in 1976. The whole enzyme substrate complex was considered but while the energy and charge distribution of the atoms directly involved in the reaction are treated quantum mechanically, the rest of atoms, including the surrounding solvent, are represented by classical forces. The same authors also implemented a force field (FF) in a computer code that was the seed of some of the more popular programs in today's computational biology, namely CHARMM (Chemistry at HARvard Molecular Mechanics, developed by Martin Karplus' group), AMBER (Assisted Model Building with

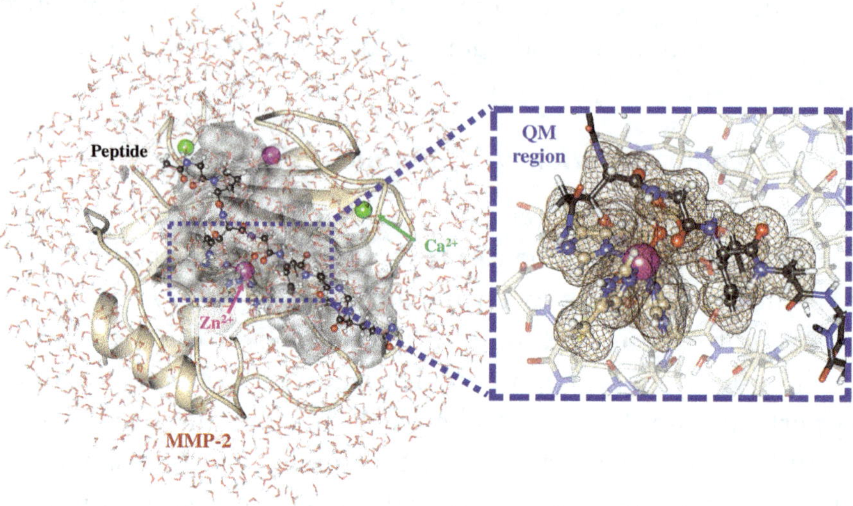

Fig. 2.3.   Ribbon model of the catalytic domain of the MMP-2 matalloen-zyme. The inset shows the active site in the complex formed with a peptide substrate.

Energy Refinement, developed by Peter Kollman's group) and GROMOS (GROningen MOlecular Simulation, developed by Wilfred van Gunsteren's group).

Figure 2.3 illustrates the QM/MM approach for the case of the complex (2665 atoms) formed by the interaction of the MMP-2 metalloenzyme (ribbon model) with a peptide substrate (ball-and-stick model). The structure was taken from a snapshot of a MD simulation using QM/MM multiscale modeling. The QM region (100 atoms) consists of the catalytic active MMP-2 residues and the peptide linkage. The MM region includes 1700 water molecules that play a key role in the simulation. The inset in the figure shows the computed charge density embedding the QM region.

The general form of a FF can be written as:

$$V(\mathbf{r}_1, \mathbf{r}_2, \ldots, \mathbf{r}_N) = V_{\text{non-bonded}} + V_{\text{bonded}} + V_{\text{solvation}} + V_{\text{water}},$$

$$(2.23)$$

with appropriate analytical expressions for the terms appearing on the right-hand side of Eq. (2.23). All of them have in common their functional simplicity to minimize computational costs. Thus, for example, usually, $V_{\text{non-bonded}}$ adopts a form like:

$$V_{\text{non-bonded}} = \sum_i \sum_{j>i} \left[\frac{q_i q_j}{4\pi\epsilon_0\epsilon_r r_{ij}} + \frac{A_{ij}}{r_{ij}^{12}} - \frac{B_{ij}}{r_{ij}^{6}} \right], \qquad (2.24)$$

a combined Coulomb + Lennard-Jones potential where q_i are atomic charges, A_{ij} and B_{ij} are fitting parameters, and r_{ij} are inter-atomic distances.

Although the parameters used to define every term in Eq. (2.24) usually have an empirical or even semi-empirical origin, parametrizations based on first principles (QM) calculations have also been very popular. A good example is provided by the so-called Matsuoka–Clementi–Yoshimine (MCY) potential for water–water interactions (V_{water} term in Eq. (2.23)), developed in the 1970s at the IBM Research Laboratories in San Jose, California.

In 1985, Roberto Car and Michele Parrinello proposed an alternative approach called *ab initio* (first principles) MD in which Newton's equations (see Eq. (2.25)) are solved for the nuclei dynamics, while the Schrödinger equation, within the so-called density functional theory (DFT) formalism, is used to describe the electronic motion. DFT methodology implies a considerable simplification in the Schrödinger equation by introducing a minimum of empiricism. The *ab initio* MD technique is a multiscale modeling in which the inter-atomic forces are computed on-the-fly by using DFT, while the nuclei propagate obeying classical mechanics. Despite its high efficiency, it represents a much more time consuming computing approach than the QM/MM methods.

Computer simulations represent a theoretical tool that allows the connection between the microscopic and macroscopic worlds through the use of statistical mechanics. One starts from the microscopic components of matter, namely the atoms (nuclei and electrons), adopting a specific geometrical disposition, as well as from the laws governing their mutual interactions. Then, statistical mechanics makes use of that information to generate theoretical predictions for the macroscopic properties of experimental interest.

Let us focus on a rather common macroscopic system: A given volume (V) of water at a given temperature (T), containing a given number of molecules (N).

From a dynamic microscopic viewpoint, the huge amount of water molecules present is in continuous motion. When we measure a given property $P(\mathbf{r}^N, \mathbf{p}^N, t)$, we get an average value $\langle P \rangle_{\text{time}}$ that corresponds to the contributions from the instantaneous values adopted by property P at the different points in the phase space (space and momentum coordinates) as the system evolves in time during the measurement process.

From a static microscopic viewpoint, an averaged value for P, $\langle P \rangle_{\text{ensemble}}$ could also be obtained by performing what is called an ensemble average over all possible quantum states of the system under the assumption that every quantum state with a given energy result is equally likely to be occupied.

While in the so-called molecular dynamics (MD) approach, one estimates the time averaged property, the Monte Carlo (MC) technique yields the corresponding ensemble averaged values. It can be shown that under the so-called ergodic hypothesis, namely, since every accessible point in a configuration space can be reached in a finite number of MC steps from any other point, both averages should be equivalent and represent the observed value of the property considered.

The MD approach has the great advantage of rendering not only the static properties of the system under study (the ones obtained from a MC simulation) but also its dynamic properties. Some relevant static (thermodynamic) properties are the radial distribution function (a microscopic property that can be experimentally determined) and temperature, pressure or heat capacity (macroscopic). Among the most important dynamic properties, we can mention the time correlation functions (microscopic) and viscosity or thermal conductivity (macroscopic). Appropriate sample techniques (using appropriate Boltzmann weighting) are required to estimate thermodynamic functions like entropy, or the Helmholtz or Gibbs free energies.

Simulations, particularly the MC technique, partly emerged from the wartime scientific research at some of the United States Department of Energy's National Laboratories, where the calculation of nuclear cross-sections resulted crucially in the development of thermonuclear weapons. The very first calculations were published in the 1950s.

In a MD simulation, we start from a N particle system with a given initial geometrical disposition and then we solve Newton's equations of motion that are assumed to be valid in the microscopic world (motions of atoms are treated in a classical way):

$$m_i \frac{d^2 \mathbf{r}_i}{dt^2} = -\nabla_i V(\mathbf{r}_1, \mathbf{r}_2, \ldots \mathbf{r}_N) \quad i = 1, 2, \ldots, N, \qquad (2.25)$$

where m_i is the mass and $\mathbf{r}_i$ is the position of the i-th particle. $V(\mathbf{r}_1, \mathbf{r}_2, \ldots, \mathbf{r}_N)$ represents the potential at the instant t.

The problem is very similar to that dealt with by Pierre-Simon Laplace more than two centuries ago in his celebrated *Mécanique Céleste*. In the present case, the main difference is that the electromagnetic forces — the relevant ones at the microscopic level — replace the gravitational force governing the macroscopic (celestial) world.

There are very efficient numerical algorithms that, coupled with appropriate computational resources, allow us today to carry out MD simulations on very large systems (10^4–10^6 atoms) and timescales beyond the μs limit.

We start from a given initial configuration by assigning initial positions and velocities to all particles in the system and then we proceed in solving Eq. (2.25) by employing finite difference methods to get a full trajectory $\mathbf{r}_i(t)$ in a number of time steps. Usually, several shorter trajectories under different initial conditions will provide better understanding of the system under study than a single long trajectory.

2.2. Some Basic Ideas on Covalent Bonding

Suppose you have two spheres flying freely and randomly, colliding with each other and with the walls from time to time.

In the beginning, the two spheres are moving independently, but once they approach each other there is some kind of force, like "glue", that binds them together. We shall discuss the nature of this "glue". This is not the regular type of glue that you are all familiar with, but rather "microscopic glue".

Now consider two hydrogen atoms: When they come closer to each other they form a hydrogen molecule that we shall denote as H_2. You can see that once they are together they stay together, practically forever. As if there is some glue holding them together. Of course, there is no such glue that binds the two atoms. The glue that you are familiar with exists in the macroscopic world, whereas in the microscopic world, glue exists only metaphorically. What binds these two atoms is a chemical bond, which can be described as very powerful *molecular glue*.

Let us consider the hydrogen molecule-ion H_2^+, which consist of two hydrogen atoms and one electron. It can be shown that

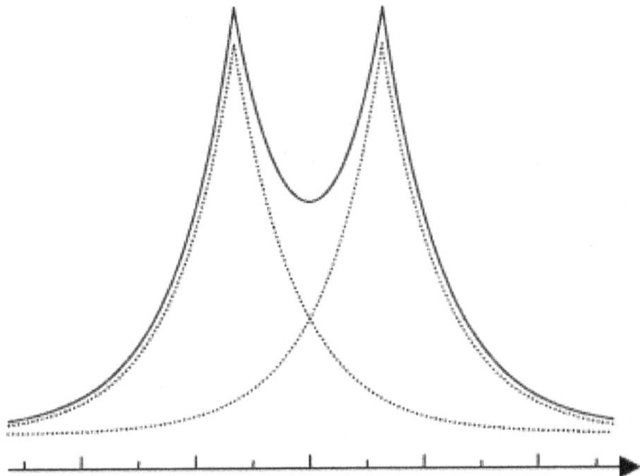

Fig. 2.4.   The ground state wave function of one electron along a 1D axis combining the centers of the positive protons attracting the common electron.

an approximate wave function, solution of the time-independent Schrödinger equation (2.18) for H_2^+, is $\Psi_\pm = N[\Psi_1(A) \pm \Psi_2(B)]$, where $\Psi_1(A)$ and $\Psi_2(B)$ are the solution of Eq. (2.18) for an hydrogen atom. The probability density of the Ψ_+ wave function, $|\Psi_+|^2$, will be $N^2[\Psi_1(A)^2 + \Psi_2(B)^2 + 2\Psi_1(A) \cdot \Psi_2(B)]$. The last term, $2\,\Psi_1(A) \cdot \Psi_2(B)$, is called overlap density, and represents an accumulation of electron density in the region between the nuclei (see Fig. 2.4)). In the case of the Ψ_- wave function, we find a negative contribution $-2\,\Psi_1(A) \cdot \Psi_2(B)$, which reduce the probability of finding the electron between the nuclei. We say that while Ψ_+ is a bonding orbital, Ψ_- is an antibonding orbital. When the electron of the H_2^+ molecule-ion occupies Ψ_+, the energy becomes lower (more stable) than that of the $H + H^+$ system. On the contrary, when Ψ_- is the molecular orbital occupied, the energy becomes greater (less stable) than that of the $H + H^+$ system.

The term $2\,\Psi_1(A) \cdot \Psi_2(B)$ represents the quantum mechanical "glue" that keeps the system H_2^+ united.

Atoms are made up of a nucleus and electrons, which may be represented by a cloud that surrounds the nucleus. The cloud represents the probability of finding the electron at that point (remember the function $|\Psi|^2$ obtained from the solution of the Schrödinger equation mentioned in the preceding section). Within this cloud, there are layers or shells of electrons. The first electron shell (called "*s*" shell) contains one orbital, which can accommodate up to two electrons. We define an "orbital" as that region of the space with a very high probability of holding two electrons that differ from each other in what is called the "spin". We can imagine the electron with spin α as a clockwise rotating electron around itself, while the β spin corresponds to another electron that is rotating counterclockwise. Energy arguments arising from the solution of the Schrödinger equation show that when the shell is complete, it is stable (lower energy E) and less likely to react.

The second electron shell can contain up to 8 electrons with 4 different orbitals. Like the first shell, it is most stable when it is complete, in this case with eight electrons.

The hydrogen (meaning "water creating" in Greek) atom, the smallest one, contains only one electron. That means there is only one vacancy in its first (and single) electron shell. Since hydrogen is more stable with a complete electron shell (two electrons), it gains stability by sharing an electron with another hydrogen atom attracted to it. The system formed by the two hydrogen atoms is stable. The two atoms are said to be covalently bonded. Figure 2.5 symbolically shows in an oversimplified manner the state in which two hydrogen atoms are covalently bonded. Both atoms share the two electrons.

The great chemist Gilbert N. Lewis (1875–1946), who made breakthrough contributions in thermodynamics, general chemistry, quantum mechanics and in many other fields, introduced

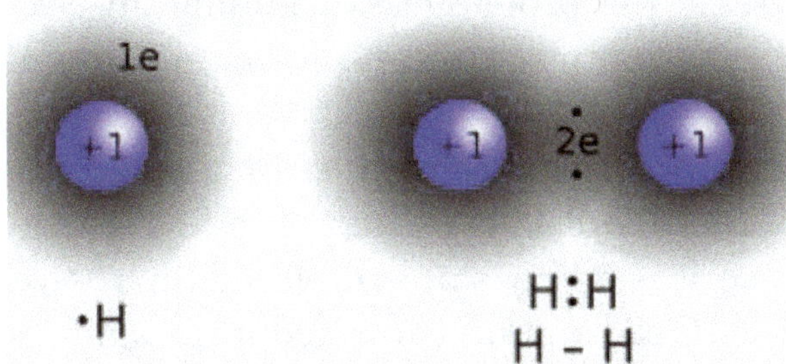

Fig. 2.5. Two hydrogen atoms form a hydrogen molecule by sharing a pair of electrons.

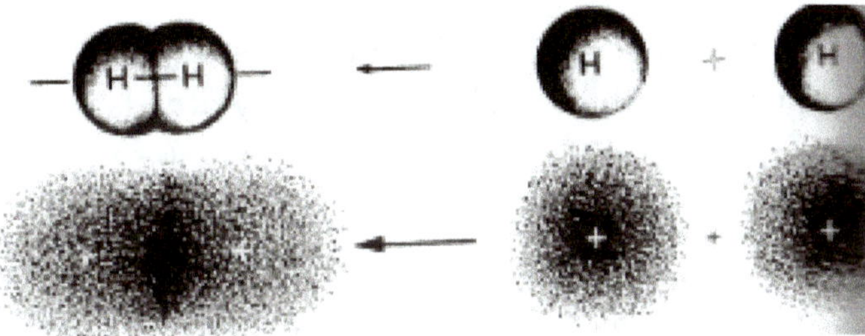

Fig. 2.6. Electron cloud of separate hydrogen atoms (right) and hydrogen molecule (left) as a combination of those two atoms. The denser the region in between the atoms of the hydrogen molecule, the higher the probability of finding electrons there. It can be seen that the denser region corresponds to that where the two electrons are shared by the two atoms, thus forming a covalent bond.

the concept of covalent bonding as that of consisting a shared pair of electrons.

The above plotting is schematic and might be misleading. An electron cannot be described by a single point. The electrons are found with great probability in between the hydrogen atoms but they can also be found, although with less probability, elsewhere. The probability of finding the electron in any location is represented by a cloud of different levels of darkness (see Fig. 2.6). The degree of darkness in any location of that cloud represents the probability of finding an electron there. That is the quantum mechanical information contained in the function $|\Psi|^2$.

A covalent bond is extremely strong as expressed by the energy needed to break it. Water bonding energy is about 459 kJ/mol, so it is a very stable molecule that is very difficult to decompose. Once a bond is formed it does not break under normal conditions.

2.3. Historical Survey About Water Composition

Let us take a breather and survey the milestones of the scientific journey that led to the conclusions that water is composed of molecules and that the water molecule can be represented as H_2O. Saying that water is just H_2O is certainly right but it is a rather simplified statement, as we are going to explain.

The concept of molecules as constituents of matter started with John Dalton (1766–1824), an English schoolteacher and amateur chemist. He realized that when substances combine to form a new one, the constituents combine in a precise ratio of masses. For example, 1 gram of hydrogen combines with 8 grams of oxygen to form 9 grams of water. This is not a self-evident conclusion according to our intuition and experience — think about cooking. However, following with the water example, 1 gram of hydrogen and 9 grams of oxygen form the same 9 grams of water, with 1 gram of uncombined oxygen left over.

Dalton concluded that the simplest explanation for this phenomenon was that oxygen and hydrogen consisted of small, indivisible particles (atoms) that combined to form a unit of water. He called this combination of atoms a *molecule*. We now know that a water molecule consists of two hydrogen atoms and one oxygen atom, but Dalton assumed that the pairing was probably one to one, meaning that an oxygen atom had eight times the mass of a hydrogen atom. He was similarly wrong for other compounds, but his atomic hypothesis was a powerful idea. Any extra atoms present had nothing to combine with, and thus remained in their free state.

On the other hand, Joseph Louis Gay-Lussac (1778–1850), a French chemist, discovered experimentally in 1808 what is now called the *Gay-Lussac's law*: When gases react chemically, the ratio between volumes of the reactants and the products is an integer number. For example, two volumes of hydrogen and one volume of oxygen combine to form two volumes of gaseous water. This is obviously similar to Dalton's results about the ratio of masses in chemical reactions, but neither Gay-Lussac nor Dalton made use of their conclusions in support of the atomic theory.

In Italy, Amedeo Avogadro (1776–1856), see Fig. 2.7, developed an eponymous hypothesis known today as *Avogadro's law* by combining the results of Dalton and Guy-Lussac: Equal volumes of gas (at the same temperature and pressure) contain the same number of particles (atoms or molecules). If that statement

Fig. 2.7. Amadeo Avogadro.

is true, Dalton's water molecule should necessarily be composed of two atoms of hydrogen for every oxygen atom since that was the ratio of volumes found by Guy-Lussac. This means that an oxygen atom had 16 times the mass of a hydrogen atom, because the two volumes of a hydrogen atom had 1/8 the mass of one volume of an oxygen atom.

The reaction that creates the water molecule produces two volumes. Three atoms combine into one molecule, so it would seem that the water vapor should have 1/3 the volume of the original gases, not 2/3. Avogadro shrewdly concluded that the original oxygen and hydrogen atoms must exist not as individual atoms, but as *pairs* of atoms combined into a single molecule (that is, H_2 and O_2). To see how this solves the problem, let us count the atoms involved in the reaction. If A is the number of particles in a unit volume, then the starting number of oxygen atoms is $A \times$ (1 volume) $\times$ 2atoms/molecule $= 2A$. The starting number of hydrogen atoms is $A \times$ (2 volumes) $\times$ 2 atoms/molecule $= 4A$. These oxygen and hydrogen atoms combine to form $2A$ molecules of H_2O, which are 2 volumes.

The constant A is directly related to the so-called *Avogadro's constant*, which we now know is 6.023×10^{23} particles per mole. Remember that a mole is a unit for amount of substance in the International System of Units.

Let us now explain something very important that is valid only in the microscopic world.

Let us focus on a system formed by hydrogen atoms filling a box. Atoms like to be free to wander around and visit all the points in the box. The higher the temperature is, the larger the speed of the flight in between the collisions. Based on this, we can say figuratively, that their tendency to be free is larger. However, if a strong bond, in our case a chemical bond, binds the two atoms together, then the pair of atoms loses some of their individual freedom. They can still wander about in the box, but now as a pair, and not as two separate individuals. It should be recalled that every single atom has three degrees of freedom (we need three coordinates to fix its position in the space). That makes six degrees of freedom for two independent atoms. However, when two atoms form a (diatomic) molecule, we only need to know five coordinates, because the sixth one can be estimated from the formula giving the (known) distance between the two atoms (distance $= [(x_2 - x_1)^2 + (y_2 - y_1)^2 + (z_2 - z_1)^2]^{1/2}$). Therefore, the diatomic system has lost one degree of freedom.

Let us to get back to liquid water molecules, which are so different from gas water molecules.

As will be shown in next few sections, it is oversimplified and incomplete to state that water is just H_2O. An individual molecule of water (H_2O) does not have any of the observable properties associated with bulk water. Should you break apart the bonds in water you will get a ratio of hydrogen atoms to oxygen atoms that is roughly 2:1; a lot of this will be from H_2O, a fair amount from other group of atoms — negative OH^- (missing hydrogen), a fair amount from H^+ etc., and it is the overall interaction of all these species, not only H_2O, that give us the correct picture of what we call water. Consequently, the properties of water arise not only from the structure of a single molecule but also from the association of a huge number of entities moving around in what we call a water bulk.

In 1883, Svante A. Arrhenius defended his thesis at the University of Uppsala. It contained the results of experimental studies on the conductivity of dilute solutions. In one of the chapters in his thesis, Arrhenius developed his theory of ionization. He concluded that current could be carried in solution, in the absence of applied potential, by "active agents" called ions that existed at all times in the solution. Although his theory was useful in explaining a number of phenomena like conductivity, osmotic pressure and vapor pressure, it left unexplained some crucial facts, for example, that once separated, one should expect ions to come back together again. His examiners passed the thesis, but with a rather poor evaluation.

Arrhenius contacted, among others, Jacobus van't Hoff and Wilhelm Ostwald, and sent them his work. Both responded favorably. Following further work with Ostwald, Arrhenius finally published his theory in the first volume of *Zeitschrift für physikalische Chemie*, the world's first journal devoted solely to physical chemistry. We will provide further details on dissociation in the next few sections.

It is interesting to remark here that experimentally supported models for the hydrated proton atom have specified hydration shells of 2 water molecules, 4 water molecules, and large clusters of up to 20 water molecules ($H_5O_2^+$, $H_9O_4^+$, $H_{41}O_{20}^+$). There is no experimental evidence supporting the existence of the bare proton (H^+). E. S. Stoyanova and co-workers have recently confirmed the existence of H_3O^+ in aqueous solution in a very limited set of circumstances. A better representation for the hydrated proton has been suggested to be $H^+(aq)$.

The only thing we can say about a glass of water is that the average ratio of atoms in the glass is 2H:1O and that it has the macroscopic properties of water, as will be described later. Moreover, water molecules are arranged in tiny structures, or microstructures, depending on how far water molecules are from the interphase air–water or from the walls of the glass. In the case of a solution containing biological molecules, water microstructures will depend on the position they occupy regarding the surface of the biological molecule as a protein, DNA, or cell's membrane. It can even be inside a macromolecule. This is due to the specific interaction between water molecules and the biological molecules.

Chemical analysis and experiments deal with water not as individual molecules but in macroscopic quantities. The molecular structure and the interactions between molecules are the basis of understanding the unusual water properties, as will be analyzed in further sections. All the typical observable properties of water — its density, its boiling and freezing points, its utility as a solvent, etc. — are dependent not only upon the isolated structure H_2O but also on the interactions among the water molecules and between water and various solutes in the solution.

2.4. Molecular Structure of Water

Probably everybody knows that the chemical formula of water molecule is H_2O, which means two hydrogen atoms are covalently bonded to a central oxygen atom. This also means that the oxygen atom shares a pair of electrons with each hydrogen atom. This structural information, as well as the electronic distribution commented below, arises from the analysis of the wave function $\Psi(r_1, r_2, \ldots, r_N)$, which is the solution of the Schrödinger equation for the water molecule.

The matter is complicated by the fact that not all H_2O molecules are the same; most involve the protium isotope of hydrogen (the most common isotope with one proton, one electron and no neutrons), a small number have deuterium (when you have water in which a very high percentage of the hydrogen is deuterium, you have heavy water, which is used in nuclear reactors; it is poisonous in large quantities), and a much smaller number involve tritium (which in large quantities would give you tritiated or super-heavy water, which is corrosive and radioactive). And that is not even counting the fact that the oxygen atoms can occur in isotopes 17, 18 and 19, each one resulting in molecules that behave differently. Water is an *interacting society*, not a single molecule, and it is a society of *related*

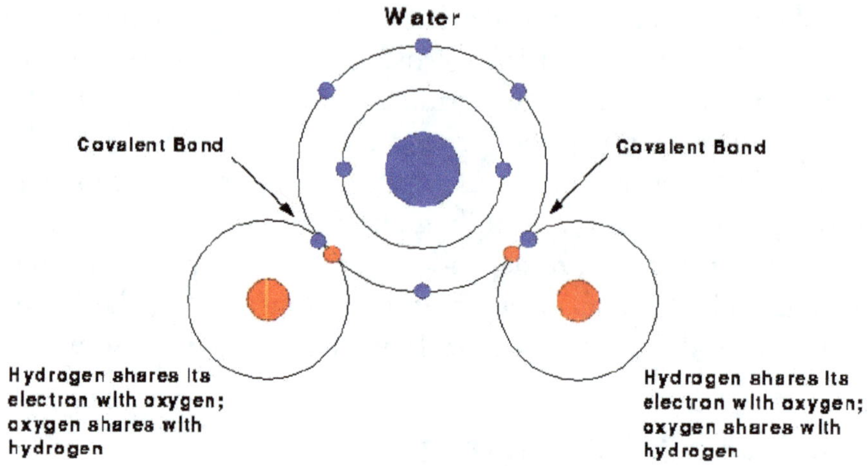

Fig. 2.8. Electronic structure of a water molecule.

molecules, not just H_2O. What in principle looks so simple is in fact a rather complicated substance.

Let us describe in a little more detail what a covalent bond is.

Only two out of the six outer-shell electrons of oxygen atoms take part in the covalent bonding (see Fig. 2.8). The remaining four electrons are organized into two non-bonding lone pairs. The two pairs surrounding the oxygen atom tend to be as far away from each other as possible in order to minimize repulsions between them. On the other hand the two non-bonding pairs push the two hydrogen atoms away from them and make the two hydrogen atoms closer to each other. The result is a distorted H–O–H angle of about 104.5° (see Fig. 2.9).

This description of a water molecule is ONLY schematic. As said, the electrons cannot be depicted as dots but rather as a cloud that represents at each point the probability of finding an electron.

The electrons are not shared equally between neighboring atoms. Atoms differ in their tendency to attract electrons towards themselves. That property is called *electro-negativity*.

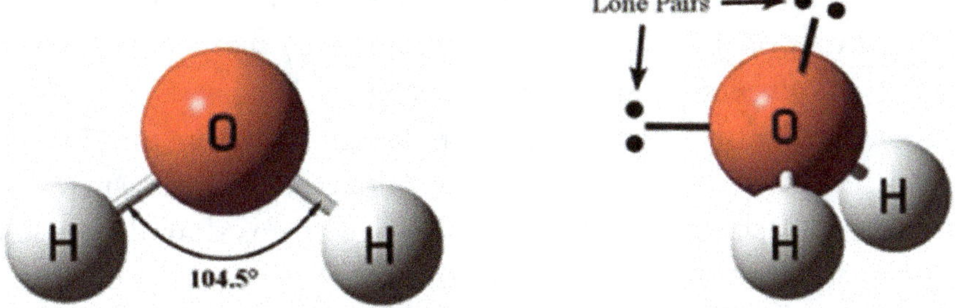

Fig. 2.9.   Water structure.

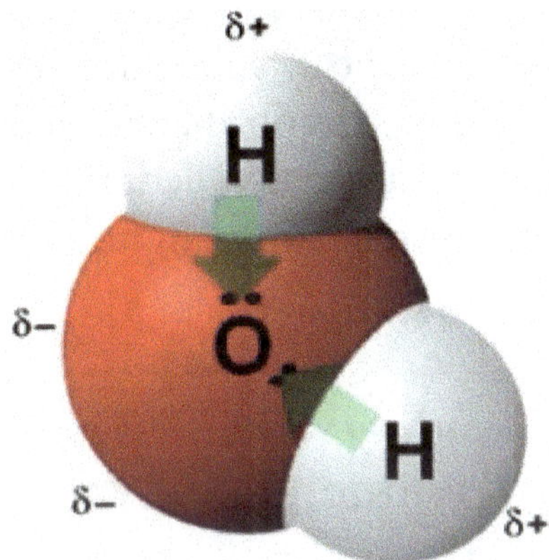

Fig. 2.10.   Polarization of water molecule.

Oxygen is more electro-negative than hydrogen, meaning that the two electrons spend more time closer to the oxygen. From a quantum mechanical point of view, the chance of finding the electron is greater in regions closer to the oxygen atom. As a result, it pulls electrons toward itself and consequently the electrons spend more time around the oxygen atom than they do around the hydrogen atoms. That causes a partial negative charge build-up on the side of the oxygen and partial positive charge around the hydrogen atoms. We say that the covalent bond is polarized (see Fig. 2.10).

The symbols δ^+ and δ^- in Fig. 2.10 indicate electrical polarity: *It is one of the most essential features of water from which most of its known properties originate.*

The nature of the bond between the oxygen atom and hydrogen atoms in the water molecule is of particular interest. In the first place, since the two different types of atoms share electrons, one can say that this is covalent bonding. However, since a part of the electron cloud is "deferred" from the hydrogen atom to the oxygen atom, they adopt opposite electric charges (δ^+ at the hydrogen, δ^- at the oxygen). The electrostatic attraction arising in this case is typical for what is called an ionic bonding, namely, the kind of bonding present in substances like NaCl in which each ion (Cl^- or Na^+) is six-coordinated by ions of the opposite sign, thus showing a local octahedral structure. Therefore, in the case of water molecule, one can talk about a mixed ionic and covalent bonding.

2.5. Hydrogen-Bonding, Loving (Hydrophilic) and Hating (Hydrophobic) Water Molecules

Next, we discuss the *intermolecular interactions forces* between water molecules. As we shall see the most important part water–water interaction is the hydrogen bond (HB) (see Fig. 2.11).

The fact that water molecules exhibit polarity has important implications in determining its interaction with other solute molecules concerns. The solute molecules can basically be classified into two groups based on whether or not they can "associate" by means of hydrogen-bonding with water. If they can, they are called hydrophilic, which means "water loving". Such substances can be dissolved in water; a good example is sugar.

In contrast to hydrophilic molecules, hydrophobic or "water fearing" molecules cannot form a HB with water and are less soluble in water; a good example is oil.

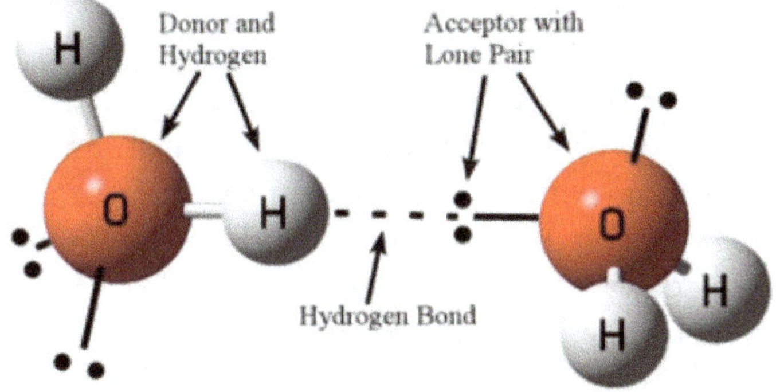

Fig. 2.11.   Hydrogen bonding.

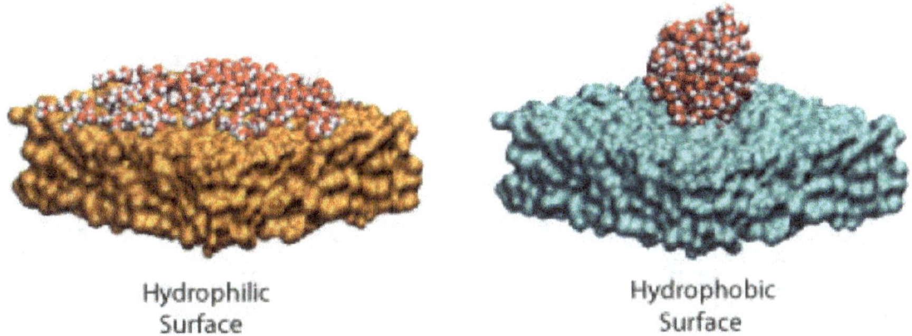

Fig. 2.12.   Water drop in contact with a hydrophilic (left) and hydrophobic (right) surface. Reproduced with permission from T. A. Isgro, M. Sotomayor and E. Cruz-Chu.

A practical consequence of the hydrophilic and hydrophobic interactions involving water is that the type of wetness caused to any given surface (by a rain drop) is determined by the hydrophobic or hydrophilic character of that surface. Over a hydrophilic surface, the water droplet spreads. Over a hydrophobic surface, it tries to avoid contact and forms a bubble. The angle between the drop smeared on the surface is a measure of the degree of hydrophilic character of the surface (see Fig. 2.12).

Hydrophilic molecular "associations" are mostly due to the HB. In 2011, the International Union of Pure and Applied Chemistry (IUPAC) defined the HB as "an attractive interaction between a hydrogen atom from a molecule or a molecular fragment $X-H$, in which X is more electro-negative than H, and an atom or group of atoms in the same or a different molecule, in which there is evidence of bond formation".

A typical hydrogen bond may be depicted as $X-H\cdots Y-Z$, where the three dots denote the bond. $X-H$ represents the hydrogen bond donor. The acceptor may be an atom or an ion Y, or a fragment or a molecule $Y-Z$, where Y is bonded to Z. In some cases, like in water, X and Y are identical (oxygen atom). The acceptor is an electron rich region such as, but not limited to, a lone pair of Y (in water, a lone pair of the oxygen atom) or π-bonded pair of $Y-Z$.

To sum up: Non-polar molecules that are said to "hate" water molecules are called hydrophobic. Molecules forming hydrogen bonds with the water molecule are said to be hydrophilic. This property of water was important for the evolution of life. Hydrophobic interaction plays an important role in the formation of the lipid bilayer of the cell membrane and the folding of proteins and nucleic acids. Until recently it was believed that hydrophobic interaction was essential for the existence of life. Hydrophobic and hydrophilic interactions will be analyzed and discussed in greater detail in Chapter 4.

Hydrogen bonds are much weaker than covalent bonds. Indeed, an $O-H$ covalent bond in a water molecule has a bond energy of about 118 kcal/mol, while the $H-O\cdots H-O$ hydrogen bonding in water is more or less 5 kcal/mol. However, when a large number of hydrogen bonds act in unison (cooperative hydrogen-bonding), they make a strong contributory effect. This is the case in water.

Therefore, water is not merely a collection of individual H_2O molecules. Instead, water in the liquid state is characterized by a specific structure.

We cannot single out a specific microstructure of water molecules as the correct one. Microstructures depend on how far water molecules are from the surface of a biological molecule as a protein, DNA, cell's membrane, and maybe inside a macromolecule. This crucial point will be analyzed in Chapter 5.

2.6. Dissociation of Ionic Compounds

Ionic bonding is usually observed when atoms with large differences in their tendencies lose or gain electrons to form a compound. For example, it happens in compounds containing a reactive metal and a nonmetal atom. This is the case of NaCl. The metal atom Na, with a low ionization energy (the energy required for Na to lose an electron giving the Na^+ ion), loses its valence electron (Na has 11 electrons: $1s^2 2s^2 2p^6 3s^1$; inner shells 1s, 2s and 2p are called "core" shells, while the outer shell 3s is called the "valence" shell), whereas the non-metal atom Cl, with a highly negative electron affinity (energy change accompanying the addition of one electron), gains the electron giving the Cl^- anion. The electrostatic attraction between the Na^+ and Cl^- ions draws them into a 3D array, thus forming an ionic compound (NaCl) in which each ion is surrounded by six ions of opposite charge occupying octahedral sites (see Fig. 2.13). It is important to stress that ionic compounds are neutral, i.e., the number of sodium cations and chloride anions is the same in NaCl crystal.

When ionic compounds dissolve in water, their ions become separated. This process is called dissociation.

When the crystal is placed in water, the polarized hydrogen atoms in H_2O ($H^{+\delta}$) interact with the chloride anions (Cl^-)

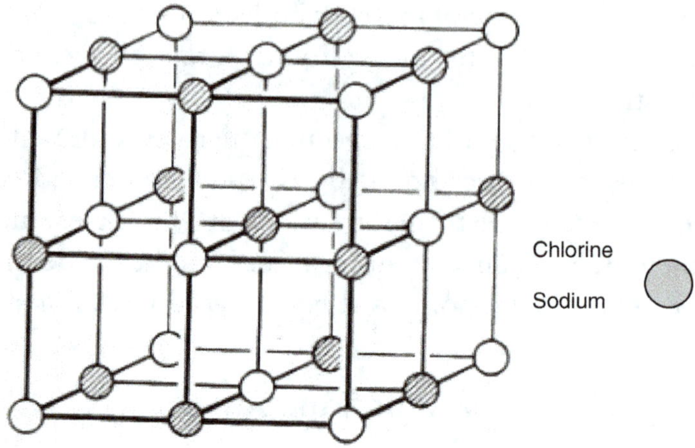

Fig. 2.13.   Lattice structure of NaCl.

Fig. 2.14.   NaCl crystal is dissolved in water.

while the polarized oxygen atoms in H_2O ($O^{-\delta}$) do the same with the sodium cations (Na^+). As a consequence, the crystal melts little by little (see Fig. 2.14).

When the crystal is completely dissolved, each sodium or chloride ion is surrounded by water molecules (ionic solvation). We say that ions are stabilized by the solvent (water) in solution. Indeed, the attractive electrostatic forces between the ions and

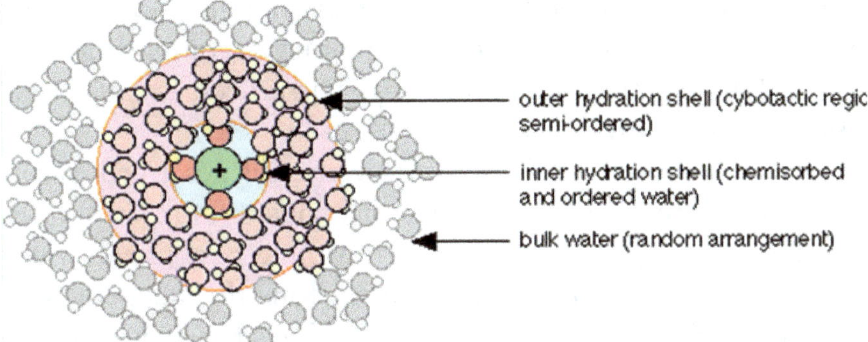

Fig. 2.15. The first coordination sphere is ordered due to the strong electric field. The second hydration shell is less ordered because of competition between the electric field of the ion and water–water HB. In the outer region we have a semi-order in between an ordered and random beginning to resemble the degree of order in the bulk water.

the water molecules make the system (solution) more stable than the crystal and bulk water separately.

2.7. Hydration: Water in Ionic Hydration Shells

Water molecules interact strongly with *ions*, which are electrically-charged atoms or molecules. Dissolution of ordinary salt (NaCl) in water yields a solution containing the ions Na^+ and Cl^-. Owing to its high polarity, the H_2O molecules closest to the dissolved ion are strongly attached to it, forming what is known as the *inner* or *primary hydration shell*. Positively-charged ions such as Na^+ attract the negative (oxygen) ends of the H_2O molecules, as shown in Fig. 2.15.

2.8. Dissociation of Covalent Bond: Water Dissociation

An interesting property of water is that two neutral molecules H_2O can interact, giving rise to two ionic species (H_3O^+,

OH$^-$). Unlike ionic compounds, such as sodium chloride, they are not ionized before they dissociate; they accomplish ionization and dissociation simultaneously. The process is called self-dissociation or autoionization.

The mechanism for the self-dissociation is described as follows: There are mainly two kinds of bonds in bulk water. On one hand, the two hydrogen atoms forming part of the water molecule are bonded to the oxygen atom through covalent unions (see Fig. 2.8). On the other hand, each water molecule interacts with the surrounding neighbors through hydrogen-bonding (see Fig. 2.16).

The water molecules are oriented in such a way that the partially negative charged oxygen faces partially positive charged hydrogen atoms in order to favor hydrogen bonding (see Fig. 2.17).

Random fluctuations in molecular motions occasionally produce a partial rupture of an oxygen–hydrogen covalent bond in a given water molecule. The forces associated to hydrogen-bonding with a second water molecule make the hydrogen atom

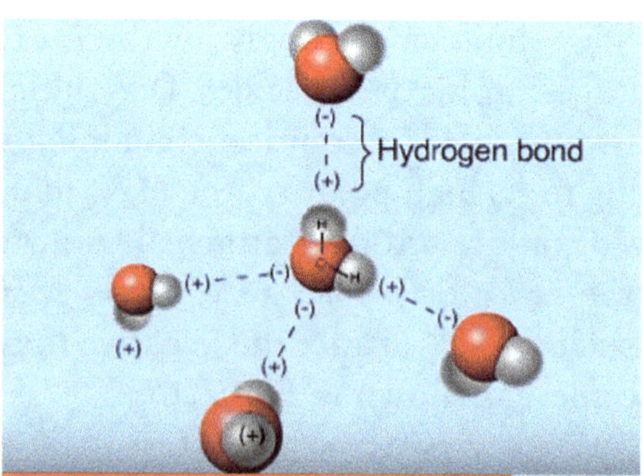

Fig. 2.16. Two existing bonding types exist in water: Covalent (see Fig. 2.8) and hydrogen-bonding (see Fig. 2.16). In a covalent bond there is electron sharing by both atoms. In a hydrogen bond there is mainly electrostatic attraction between opposite charges.

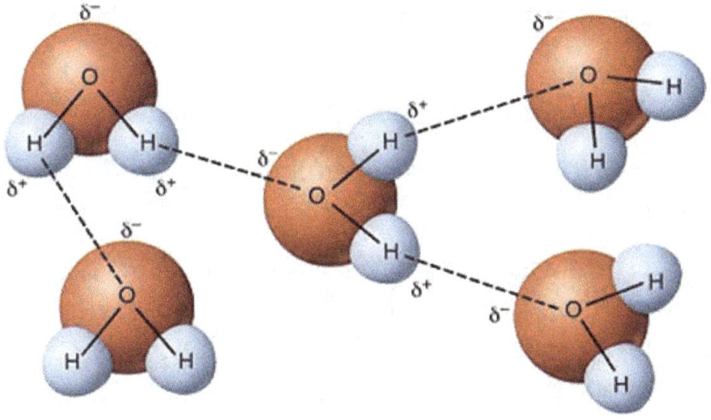

Fig. 2.17. Orientation of water molecules ready for dissociation.

of the first molecule to share one of its two electron pairs of the oxygen in the second water molecule. Thus finally, the first water molecule becomes an (hydroxide) OH^- ion (it retained the electron pair of the covalent bonding corresponding to the leaving hydrogen), while the second water molecule becomes a (hydronium) H_3O^+ ion.

From a chemical point of view, the process described above (self-dissociation) can be written as:

$$2H_2O \leftrightharpoons H_3O^+ + OH^-,$$

or, in a more detailed description:

$$\overset{H}{\underset{H}{\diagdown}}\ddot{O}\!: \qquad H-\ddot{O}\underset{\diagdown H}{} \;\rightleftharpoons\; \left(\overset{H}{\underset{H}{\diagdown}}\ddot{O}-H\right)^{+} + \left(\!:\!\ddot{O}\underset{\diagdown H}{}\right)^{-}$$

It should be noted that the process reaches an equilibrium in which the rate of ion production equals the rate of neutral water formation. However, the equilibrium is clearly shifted towards neutral water. That is to say, the presence of H_3O^+ and OH^-

ions is minimal. We will come back to this point in the next few sections.

2.8.1. *Concentration of Dissolved Substance in Terms of Moles*

A main question arises now: What is the concentration of each of these ions when a state of equilibrium is reached?

The International Union of Pure and Applied Chemistry (IUPAC) defines the mole as the amount of substance of a system that contains as many elementary entities as there are atoms in 0.012 kilograms of carbon-12 (^{12}C is the more abundant isotope of carbon, with 6 neutrons and 6 protons. Another rather popular isotope of carbon is ^{14}C with 8 neutrons and 6 protons. It is a radioactive isotope of carbon whose preference in organic materials is the basis of the radiocarbon dating method). In that definition, "elementary entities" mean atoms, molecules, ions, electrons, other particles or specified groups of such particles. The mole is one among the seven base units of the International System of Units (SI).

Let us now define the atomic mass unit (amu) as 1/12 the mass of a ^{12}C atom. In order to find the equivalence between the new mass unit (amu) and the SI unit of mass (1 kg), let us recall that one ^{12}C atom contains 6 protons, 6 neutrons and 6 electrons (a total of $2.0087 \cdot 10^{-26}$ kg). On the other hand, the stabilization energy released in the formation of ^{12}C nucleus is $92.16\,MeV$ ($1\,MeV = 10^6\,eV$, and $1\,eV = 1.6022 \cdot 10^{-22}\,kJ$), which according to Einstein's relation $E = mc^2$, represents a mass loss of $1.6429 \cdot 10^{-28}$ kg. The above data leads to a mass of $1.9923 \cdot 10^{-26}$ kg for the ^{12}C atom. That is to say, 1 amu is $1.66 \cdot 10^{-27}$ kg.

The relative atomic mass or atomic weight is defined as the quotient between the atomic mass and the atomic mass unit

(1/12 times the mass of a ^{12}C atom). Thus, in the case of ^{12}C, the atomic weight will be $1.9923 \cdot 10^{-26}$ kg$/1.66 \cdot 10^{-27}$ kg $= 12$ atomic units. In a similar way, one can talk of relative molecular mass or molecular weight.

This important result can be generalized as follows: The Avogadro's number of atoms of any element has a mass in grams identical to the atomic weight of the element. Indeed, let us consider as an example the nitrogen atom. Its atomic weight is 14.01. That means that the mass of the nitrogen atom is 14.01/12 times the mass of a ^{12}C atom. If instead of one atom we consider $6.02 \cdot 10^{23}$ atoms of both nitrogen and ^{12}C, the total mass of the nitrogen (now 1 mole) will be 14.01/12 times the mass of 1 mole of ^{12}C, which according to the SI definition is 12 g, i.e., 14.01 g. That is to say, the Avogadro's number of atoms of nitrogen has a mass in grams identical to the nitrogen atomic weight.

Another useful concept is that of molar mass, the mass per mole of a pure substance. For example, the molar mass for water is 18.015 g/mol (18.02 if we use data from Fig. 2.18).

The IUPAC defines the so-called Avogadro's constant (N_A) as an Avogadro's number with units of mol^{-1}. If we consider a system with N_i molecules (elementary entities; dimensionless) the amount of substance n_i (moles) is, according to the IUPAC mole definition, related to it by $N_i = n_i \cdot N_A$, where $N_A =$ Avogadro's constant $= 6.023 \cdot 10^{23}$ mol^{-1}.

Once we have introduced the concept of mole, we can now focus on how to express the concentration of a given solution.

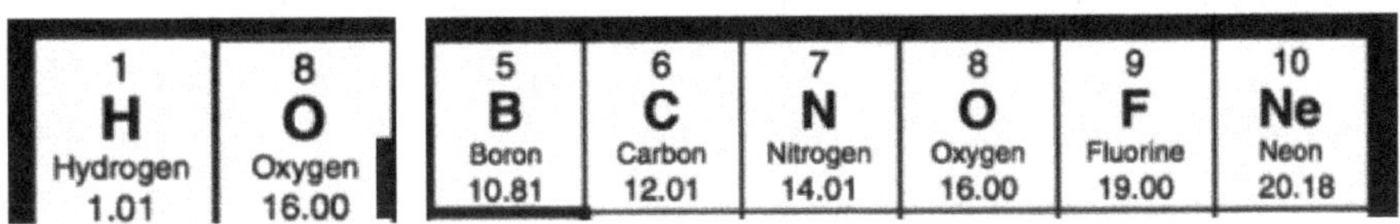

Fig. 2.18. Some elements in the Periodic Table showing their atomic weights.

We will call solute, in general, the substance in a smaller amount in the solution.

The concentration of a solution is usually expressed as the amount of solute dissolved in a given amount of solution. Perhaps the most common concentration scale employed in chemistry is the molarity (M): The number of moles of solute per liter of solution. If we compute the molarity times the molar mass, we get concentration in grams of solute per liter of solution, another very common concentration scale.

2.8.2. *Defining pH as a Measure of Acidity and Basicity*

As in any chemical equilibrium, self-dissociation is governed by an equilibrium constant:

$$K_c = \frac{[H_3O^+][OH^-]}{[H_2O]^2}$$

The notation $[a]$ means concentration of a.

The concentration of pure water $[H_2O]$ (55.35 M; readily obtained from its density at 25°C and 1 atm: $\rho = 0.997043\,\text{g} \cdot \text{cm}^{-3}$, and its molar mass: M = 18.015 g/mole) remains constant and one can include the term $[H_2O]^2$ in K_c to obtain the so-called ion product constant for water, K_w:

$$K_w = [H_3O^+][OH^-] = 10^{-14}.$$

This equation remains applicable to dilute solutions where $[H_2O]$ hardly changes.

Since the stoichiometry of the self-dissociation reaction of water indicates that $[H_3O^+] = [OH^-]$, then we get:

$$[H_3O^+] = [OH^-] = 10^{-7}\,\text{M}.$$

Let us analyze in some detail the above result. When we drink a glass of water, we are swallowing mostly neutral water molecules as well as a rather small amount of hydronium and hydroxide ions. Indeed, one liter of pure water contains 55.35 moles of neutral water molecules and 0.0000001 moles of H_3O^+ and OH^- ions. The amount of ions is almost negligible. Even so, 0.0000001 moles mean about 60,000,000,000,000,000 $(6 \cdot 10^{16})$ molecules!

When huge numbers are involved, it is usual to resort to a logarithmic scale. The Danish biochemist Søren Sørensen suggested the use of the term pH ("potential of hydrogen ion") defined as:

$$pH = -\log[H_3O^+].$$

According to that definition, we get pH = 7 for pure water. If we apply the same definition to the hydroxide concentration:

$$pOH = -\log[OH^-]$$

and therefore, for pure water pH = pOH = 7.

The application of the mathematical properties of the logarithmic function to K_w gives:

$$pK_w = pH + pOH = 14.$$

We say that at 25°C and 1 atm, any aqueous solution with pH = 7 (the pH for pure water) is a neutral solution. If pH < 7 the solution is said to be acidic and if pH > 7, it is basic.

The pH values are readily obtained in the laboratory either by using an acid-base indicator or by applying a pH meter. The acid-base indicators are chemical substances showing a different color depending on the acidity or basicity of the solution. The pH meter determines $[H_3O^+]$ by measuring the voltage difference between two electrodes immersed in the solution. Figure 2.19 presents the pH for some substances we deal with in our daily life.

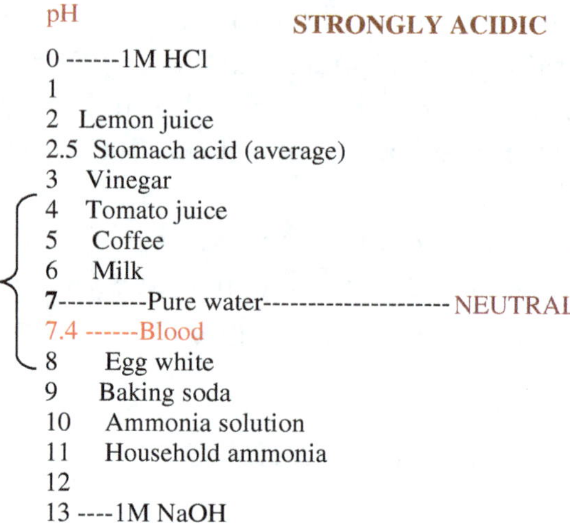

Fig. 2.19.   The pH of different substances in our daily life. Relative acid-base strength. Stronger acids (bases) have weaker conjugate bases (acids). This diagram allows us to predict the net direction of an acid-base reaction. For the ammonia hydrolysis analyzed in the text, we note that NH_4 is a stronger acid than H_2O and OH^- is a stronger base than NH_3. Therefore, the reaction will proceed to the left.

2.9. Acids and Bases

The balance between hydronium and hydroxide ions is violated in favor of one of them if other molecules are dissolved in water: Acids, bases or salts. They may be covalent, ionic or partially ionic substances, and may change considerably the concentration of hydrogen and hydroxyl ions in solution.

2.9.1. *Acids*

Most acids contain hydrogen in their formulas. An example is hydrochloric acid, HCl, containing hydrogen and chlorine with a partial polarized covalent bond between them. When HCl dissolves in water its proton (the hydrogen nucleus, i.e., the hydrogen atom without the electron) bonds to the water molecule through one of the lone pairs in oxygen, forming the H_3O^+ ion. It leaves a chlorine ion (Cl^-) in solution. Both ions become stabilized by surrounding water molecules, as we mentioned previously.

The term dissociation of HCl that one has probably learned in high school is a little bit misleading. The statement that HCl "dissociates" in water actually does not mean that the HCl molecule suddenly breaks into H^+ and Cl^-, because if that was the case, one will need to supply enormous amounts of energy (bearing in mind that the enthalpy of the formation of H^+ and Cl^- are $1536.2\,kJ \cdot mol^{-1}$ and $-234.0\,kJ \cdot mol^{-1}$, respectively, the energy cost is $1302.2\,kj \cdot mol^{-1}$). What actually happens is that the HCl reacts with water and the reaction is an exothermic reaction. That is why when one dilutes HCl in water, the reaction glass becomes slightly warmer.

The concentration of the hydronium ion may increase considerably when an acid is dissolved in water. The lower the pH the higher the concentration of hydrogen ions in the solution. As an example, if we dissolve 0.01 moles of HCl in one liter of water solution, we get a hydronium concentration of $10^{-2}\,M$, which sharply contrasts with the corresponding value in pure water ($10^{-7}\,M$). The pH changes from 7 in pure water to 2 in the solution.

2.9.2. *Bases*

A strong base can give your skin a much worse burn than an acid. A typical strong base is sodium hydroxide, NaOH. NaOH dissociates to form a sodium ion and a hydroxide ion.

$$[NaOH]_{aq} \rightarrow Na^+_{aq} + OH^-_{aq}.$$

Let us consider a $0.01\,M$ solution of NaOH in water. We mentioned in the previous section that in dilute solutions $[H_3O^+] \cdot [OH^-] = 10^{-14}$. Therefore, assuming that NaOH is fully dissociated, $[OH^-] \approx 10^{-2}$, since we neglect the $[OH^-]$ concentration arising from self-dissociation of water for being much smaller. Indeed, such a concentration is expected to be smaller that 10^{-7}, its value in pure water, because Le Châtelier's principle dictates that self-dissociation equilibrium water must shift to the formation of neutral water molecules as a consequence of the presence of OH^- ions from the dissociation of NaOH. Therefore $[H_3O^+] = 10^{-12}$ and pH $= 12$ for this solution.

2.9.3. *Strength of Acids and Bases in Water*

In previous sections, it was assumed that concentrations of hydronium and hydroxyle ions in water were that of HCl and NaOH, respectively (fully dissociation). This is because HCl and NaOH are a strong acid and base, respectively. The corresponding equilibria are virtually fully shifted toward the right (formation of hydronium or hydroxyle ions).

When the equilibrium is only partially shifted to the right, we talk about weak acids or weak bases. Let us consider, for instance, the case of acetic acid:

$$HCOOH_{(liquid)} + H_2O_{(liquid)} \leftrightarrows H_3O^+_{aq} + CH_3COO^-_{aq}.$$

The equilibrium constant for this acid dissociation is:

$$K_c = \frac{[H_3O^+] \cdot [CH_3COO^-]}{[HCOOH] \cdot [H_2O]}.$$

In the case of dilute solutions, $[H_2O]$ is so much larger than $[HCOOH]$ that it can be considered as a constant when acetic acid dissociates. We call $K_c \cdot [H_2O]$ the acid dissociation constant K_a.

$$K_a = \frac{[H_3O^+] \cdot [CH_3COO^-]}{[HCOOH]}.$$

The stronger the acid, the higher the equilibrium concentration of hydronium atoms is, and the larger the value of K_a.

The above arguments apply equivalently to base dissociations, where the base dissociation constant K_b is introduced.

> Acids and bases are cornerstones of life in our planet.
>
> We cannot avoid the concept of acids and bases even in everyday life. Nearly any liquid has acidic or basic properties, including fluids in our bodies. Water is an exception.

2.9.4. *Three Definitions of Acids and Bases*

In the preceding section, we discussed and analyzed the behavior in water solution of acids and bases. We assumed the classical definition introduced by Svante Arrhenius, where an acid is a hydrogen compound that gives rise to hydronium ions in aqueous solution. In a similar way, a base is a substance that has the OH group in its formula and dissociates in water to yield hydroxyle ions. The neutralization of acids by bases could be explained in terms of the two ions to give the neutral molecular water. The soundness of Arrhenius' theory was illustrated by the uniformity of the heats of neutralization (about $-55.9\,\mathrm{kJ/mol}$) whenever equivalent solutions of any strong acid and any strong base neutralize one another. Arrhenius' theory was generally accepted for 40 years. However, a number of limitations soon became apparent. Thus, for example, the basic character of ammonia, NH_3, hardly could be explained in terms of this theory, since no OH is present in its formula. An *ad hoc* explanation calls for the formation of the substance NH_4OH in aqueous solution, and subsequent dissociation according to the following scheme:

$$NH_{3g} + H_2O_l \rightarrow NH_4OH_{aq},$$

$$NH_4OH_{aq} \leftrightarrows NH_{4\,aq}^{+} + OH_{aq}^{-}.$$

the subscript "g" for gas and "l" for liquid.

The problem is that there is no clear evidence about the presence of the species NH_4OH_{aq} in aqueous solution.

A second, more generalized, definition was proposed in 1923 almost simultaneously by Johannes Nicolaus Brønsted and Thomas Martin Lowry. According to them, an acid is a species having a tendency to lose a proton (H^+) and a base is a species having a tendency to gain a proton. It is important to stress that the definition involves the species H^+ rather that the actual hydrogen atoms that occur in different solutions. Therefore, the new definition is independent of the solvent. According to the Brønsted–Lowry theory, it is straightforward to rationalize the basic character of ammonia:

$$NH_{3g} \text{ (Base 1)} + H_2O_l \text{ (Acid 2)}$$

$$\leftrightarrows NH_{4aq}^+ \text{ (Acid 1)} + OH_{aq}^- \text{ (Base 2)}.$$

An interesting new concept arises from the Brønsted–Lowry definition. Consideration of the above reaction towards the right-hand side leads to the conclusion that NH_3 acts as a base because it gains a proton from H_2O, which then acts as an acid. But, at the same time, consideration of the reaction towards the left-hand side shows that NH_4^+ acts as an acid giving a proton to OH^- that operates like a base gaining that proton. We can formally establish that NH_3 (base) and NH_4^+ (acid) are a conjugate acid-base pair. Every acid has a conjugate base, and every base has a conjugate acid.

The equilibrium constant for the considered reaction is:

$$K_b = \frac{[NH_4^+] \cdot [OH^-]}{[NH_3]} = 1.8 \cdot 10^{-5}.$$

$K_b < 1$ means that NH_3 is a base weaker than OH^-, and H_2O is an acid weaker than NH_4^+. The reaction proceeds to the greater extent in the direction in which the stronger species form the weaker species (to the left), see Fig. 2.19 for the acid-base strength of common conjugate acid-base pairs.

Another important feature of the Brønsted–Lowry theory is that it considers the possibility that the same species possesses both acidic and basic properties. An excellent example is water. According to the self-dissociation reaction discussed in a previous section, water acts as an acid and as a base. Water is said to be amphoteric. The amphoteric character of water is particularly important in determining its properties as a solvent for acid-base reactions.

It should be noted that every species being an acid/base according to Arrhenius' theory would behave as an acid/base in the Brønsted–Lowry sense. However, some species with no OH group in their formulas will be considered Brønsted–Lowry bases, although they do not admit their classification as Arrhenius' bases. As a prototypical example of this latter assertion we can mention the case of ammonia discussed above.

A third definition of the acid-base concept is due to the great American chemist Gilbert Newton Lewis, who in 1923 defined an acid as a species that could accept an electron pair from a base with the formation of a covalent chemical bond. The base donates the electron pair and the acid offers a vacant orbital where the electron pair is hosted. Brønsted–Lowry's and Lewis' definitions are equivalent since a molecule that can accept a proton (Brønsted–Lowry base) must have an electron pair available (Lewis base) to form the covalent bond. However, Lewis' definition of an acid does not need to involve any proton transference. Thus, typical Lewis acids like SO_2 or BF_3 are non-hydrogen containing molecules. They have vacant orbitals ready to accept a lone pair and form a new covalent bond. Lewis introduced the new concept of these nonprotic acids because it was experimentally observed that these substances exhibited the properties regarded as typical of acids, such as neutralization of bases, action on indicators and catalysis.

Therefore, Lewis' definition represents a generalization of Brønsted–Lowry's definition. To put it in Lewis' words, to restrict the group of acids to those substances that contain hydrogen would interfere as seriously with the systematic understanding of chemistry as would the restriction of the term oxidizing agent to those substances containing oxygen.

2.10. Hydrolysis of Salts

Many salts give aqueous solutions with acidic or basic properties. This is termed *hydrolysis*, and the explanation of hydrolysis reactions in classical acid-base terms was somewhat involved. In terms of the Brønsted–Lowry theory, however, hydrolysis appears to be a natural consequence of the acidic properties of cations derived from weak bases and the basic properties of anions derived from weak acids.

As we have seen in the previous section, dissociation of strong acids and bases give rise to anions and cations (Cl^-, Na^+) that become hydrated (surrounded by water molecules). No reaction with water takes place. Therefore, a solution containing only these ions must be neutral ($pH = 7$). Indeed, if we dissolve common salt ($NaCl$) in water, the resulting solution is neutral. This fact can be generalized in the following way: Salts containing an anion arising from a strong acid and a cation from a strong base yield neutral solutions.

Let us now see what happens when we dissolve a salt with the anion of a strong acid and the cation of a weak base. As an illustration of this process we will focus on NH_4Cl salt:

$$NH_4Cl_s + H_2O_l \rightarrow NH_{4\,aq}^+ + Cl_{aq}^-.$$

When added to water, ammonium chloride dissociates to release NH_4^+ and Cl^- ions that get hydrated. While the Cl^- ion remains this way because it is an anion of the strong acid HCl, the NH_4^+ cation is a weak acid, and consequently will react with water according to:

$$NH_4^+ + H_2O_l \leftrightarrows NH_{3aq} + H_3O_{aq}^+,$$

giving rise to an acidic solution.

Therefore, we arrive to the conclusion that solutions of salts containing anions of strong acids have an acid character. It is important to stress that the equilibrium of the ammonium cation hydrolysis reaction is shifted to the left, as NH_3 is a stronger base than H_2O and H_3O^+ is a stronger acid than NH_4.

In the case of solutions containing a salt with the cation of a strong base and the anion of a weak acid (for example, sodium acetate), similar considerations lead to the following equations:

$$CH_3COONa_l + H_2O_l \rightarrow Na_{aq}^+ + CH_3COOH_{aq}^-,$$

$$CH_3COO_{aq}^- + H_2O_l \leftrightarrows CH_3COOH_{aq} + OH_{aq}^-.$$

The second reaction is an equilibrium shifted to the left because OH^- is a stronger base than CH_3COO^- and CH_3COOH is a stronger acid than H_2O.

Therefore, solutions of salts containing cations of strong bases (NaOH) and anions of weak acids (HCOOH) have a basic character.

In the case of salts of weak acidic cations (like NH_4^+) and weak anionic bases (like $HCOO^-$), the reactions involved will be the following:

$$NH_4^+ + H_2O_l \leftrightarrows NH_{3aq} + H_3O_{aq}^+,$$

$$CH_3COO_{aq}^- + H_2O_l \leftrightarrows CH_3COOH_{aq} + OH_{aq}^-.$$

The final pH will be dictated by the relative values of the corresponding equilibrium constants. In this particular case, $K_a = 5.7 \cdot 10^{-10}$ and $K_b = 5.6 \cdot 10^{-10}$, the final pH will be virtually neutral (pH $\approx$ 7).

2.11. Effects of pH on Health and Environment

Our body has a number of fluids and tissues (blood serum, aqueous humor of eye, saliva, pure gastric juice, urine, tears, etc.), each with its specific pH that should vary within a rather narrow margin. Let us comment briefly on a usual chemical mechanism our body employs to keep the pH constant. We will focus on one of the fluids: Blood plasma, with pH $= 7.4$. (A blood pH out of the 7.35–7.45 range can affect our health negatively. It can be even fatal in some occasions.)

When we exercise, our body's metabolism becomes more active, producing CO_2 and H^+ in the muscles during the breakdown of glucose. These substances are removed from the muscles via the blood. Our body has two mechanisms to prevent a lowering of pH as a consequence of the increase in H^+ — buffering and excreting H^+. Let us focus on the buffering action.

A buffer solution is a solution whose pH remains essentially constant despite the addition of small amounts of acids or bases. The best way to show how buffering works is through a practical example. Let us consider one of the simplest buffers — the acetic-sodium acetate system.

Our system consists of one liter of solution, which contains one mole of acetic acid and one mole of sodium acetate at 25°C. The reactions involved are the following:

$$CH_3COONa_l + H_2O_l \rightarrow Na^+_{aq} + CH_3COOH^-_{aq},$$

$$CH_3COOH_l + H_2O_l \leftrightarrows CH_3COO^-_{aq} + H_3O^+_{aq},$$

$$CH_3COO^-_{aq} + H_2O_l \leftrightarrows CH_3COOH_{aq} + OH^-_{aq},$$

$$H_2O_l + H_2O_l \leftrightarrows H_3O^+_{aq} + OH^-_{aq}.$$

The two first equations correspond to the ionization of the weak acid and the salt. The third reaction represents the reaction with water (hydrolylis) of the weak base acetate. Finally, the last reaction is the self-dissociation of water.

It is easy to show that considering the initial concentrations of acid, $[CH_3COOH]_0$, and salt, $[CH_3COONa]_0$, and taking into account the electro-neutrality condition, $[Na^+]+[H_3O^+] = [CH_3COO^-] + [OH^-]$, the ion-product constant for water, $[H_3O^+] \cdot [OH^-] = 10^{-14}$, and the dissociation constant of the weak acid K_a, one gets:

$$K_a = \frac{[H_3O^+] \cdot [CH_3COO^-]}{[HCOOH]}$$
$$= \frac{[H_3O^+]\{[CH_3COO^-]_0 + [H_3O^+] - [OH^-]\}}{[HCOOH]_0 - [H_3O^+] + [OH^-]}.$$

Practical considerations (the acid will dissociate slightly and the anion will hydrolyze slightly. These effects will nearly cancel out each other and the concentrations of acid and anion will virtually correspond to those of the initial acid and salt, respectively) allow one to neglect, in general, $[H_3O^+]$ and $[OH^-]$ with respect to $[CH_3COONa]$ and $[CH_3COOH]$. Then, we finally obtain the expression:

$$K_a = [H_3O^+]\frac{[CH_3COO^-]_0}{[HCOOH]_0} \Rightarrow pH = pK_a + \log \frac{[CH_3COO^-]_0}{[HCOOH]_0},$$

which is the so-called Henderson–Hasselbalch equation.

What happens when a small amount of strong acid is added to the solution? In order to provide a quantitative answer to that question, let us consider a $1\,M\;CH_3COOH/1\,M\;CH_3COONa$ buffer solution ($K_a = 1.8 \cdot 10^{-5}$). According to the Henderson–Hasselbalch equation, the resulting buffer pH will be 4.74. If we add 0.1 mole of HCl to one liter of the buffer solution, the existing $[CH_3COO^-]$ will neutralize the $[H_3O^+]$ added, forming CH_3COOH. The new values of $[CH3COO^-]$ and $[CH_3COOH]$ will be 0.9 M and 1.1 M, respectively. Consequently, the new pH will be 4.65, quite close to the pH of the buffer solution. It should be noted that if the 0.1 M concentration of HCl is added to pure water, the pH would have changed from 7 to 1. This is the marvelous attenuation effect of buffer solutions! It is straightforward to confirm that a similar kind of behavior is observed when adding a strong base to the buffer solution.

Two important concepts related to buffer solutions are buffer capacity and buffer range. Buffer capacity measures the competence to resist pH change. It can be easily shown that the buffer with the highest capacity is that with identical concentrations of the weak acid and its conjugate base ($[CH_3COOH] = [CH_3COO^-]$). The buffer range is the pH range over which the buffer acts effectively. The optimum buffer range is that given by $pH = pK_a \mp 1$.

In the case of the blood plasma, one of the buffers operating to maintain the pH at 7.4 is the HCO_3^-/H_2CO_3 buffer:

$$H_2CO_{3aq} + H_2O_l \leftrightharpoons H_3O_{aq}^+ + HCO_{3aq}^- \, (K_a = 2.5 \cdot 10^{-4} \text{ at } 25°C).$$

The corresponding Henderson–Hasselbalch equation is:

$$pH = pK_a + \log \frac{[HCO_3^-]}{[H_2CO_3]}.$$

The carbonic acid decomposes in water according to:

$$H_2CO_{3aq} + H_2O_l \rightleftarrows CO_{2g} + 2H_2O_l \, (K = 5.9 \cdot 10^2 \text{ at } 25°C).$$

Then, the final equation will be:

$$pH = 6.1 + \log \frac{[HCO_3^-]}{[CO_2]},$$

where 6.1 is the temperature corrected value of $pK_a - pK$ ($=6.37$), bearing in mind that the body's temperature is about 37°C.

A normal person has a concentration of about $1.2 \cdot 10^{-3}$ M of CO_2 and 0.024 M of HCO_3^- in blood. This gives pH $= 7.4$ for blood plasma. When the body increases $[H_3O^+]$ and $[CO_2]$ after hard exercise, the ratio $[HCO_3^-]/[CO_2]$ (an adult weighing 70 kg generates about 0.1 mole of protons and 12 mole of CO_2 per day, just because of metabolism) changes: Le Châtelier's principle predicts a decrease of $[HCO_3^-]$ as a consequence of the increase in the concentration of protons, thus lowering the pH. The HCO_3^-/H_2CO_3 buffer helps keep the pH constant.

However, the blood pH of 7.4 is outside the optimal buffering range of 5.1–7.1. Therefore, other mechanisms are expected to be activated to help keep the $[HCO_3^-]/[CO_2]$ ratio constant. Thus, the lungs remove excess CO_2 from the blood, and the kidneys remove excess HCO_3^- from the body. Our body is a marvelous machine combining different mechanism, as required, to keep us in perfect shape!

2.12. Effects of pH on Rain, Oceans and Lakes

Rainwater is acidic because raindrops dissolve the atmospheric CO_2, giving rise to the following equilibrium equations:

$$CO_{2g} + H_2O_l \rightleftharpoons CO_3H_{2aq},$$

$$CO_3H_{2aq} + H_2O_l \rightleftharpoons H_3O_{aq}^+ + HCO_{3aq}^-,$$

$$HCO_{3aq}^- + H_2O_l \rightleftharpoons H_3O_{aq}^+ + CO_{3aq}^{2-}.$$

For a global atmospheric concentration of CO_2 of about 350 ppm (350 mg/L), it is easy to estimate from the equilibrium constant involved in the above equations that the final pH has a value of about 5.6. Therefore, natural rainwater is acidic. Even in the absence of anthropogenic emissions, rainwater is naturally acidic (although it can occasionally be neutral or alkaline as a consequence of contact with alkaline minerals in wind-blown dust, or with gaseous ammonia from soils). But the presence in the atmosphere of a number of pollutants like sulfuric acid (sulfur dioxide from the fossil fuel combustion dissolves in water, giving sulfurous acid, which in turn reacts with oxidant atmospheric agents like ozone, thus producing sulfuric acid), nitric acid (nitrogen oxides, NO_x, arising from combustion in car engines and electric power plants that hydrolyzes to nitric acid) or chlorhydric acid (produced by the reaction of sulfuric acid with atmospheric sodium chloride of marine origin), which makes rainwater much more acidic. As early as 1692, Robert Boyle alluded to the presence of sulfur compounds and acids in air and rain. But it was the English chemist Robert Angus Smith who introduced the term "acid rain".

Rainwater pH varies from location to location, depending on different causes, one of the most important factors being the local generation of anthropomorphic pollutants giving rise to the acid rain. Thus, data from the 1980s show that rainwater pH was about 5.3 in Spain, 4.5–4.7 in England, 5.1–5.3 in France, or a medium value of around 4.4 in North America, with remarkable peaks in Wheeling, West Virginia, where the pH reached a critical value of 1.8!

It should not be forgotten that a difference in 3.5 pH units (logarithmic scale!) means a 30,000-fold excess of proton concentration of the rainwater in West Virginia with respect to Spain. Such excess is expected to have a very negative environmental impact. The most remarkable damages incurred directly from the acidic deposition include the acidification of oceans' surface waters (lakes, rivers, etc.), damage to forests and vegetation, and finally, damage of materials and structures.

Atmospheric carbon dioxide is absorbed to a considerable extent by the ocean. It reacts with seawater to form carbonic acid, and this causes the acidity of seawater to increase, according to the chemical equations above (ocean acidification). That increase in acidity exerts a rather negative impact on the proliferation of corals or the survival of shellfish, whose shells are made of calcium ion (Ca^{2+}) with carbonate ion (CO_3^{2-}) from surrounding seawater. Should seawater contain protons from ocean acidification, analysis of the corresponding equilibrium constants involved tells us that formation of H_2CO_3 is a more favored process, making it harder for shelled marine fauna to form CO_3Ca to build their structures.

Regarding a lake's acidification, the presence of strong acids in waters containing bicarbonate ions HCO_3^- can be controlled by some extent through the CO_3H^-/CO_3H_2 buffer analyzed in a previous section. Soil around the surface water also helps to buffer the pH of the water by means of proton exchange with a number of cations like Ca^{2+} or Na^+.

The lake's acidification exerts great influence on biological material (fish and vegetation), producing in some cases fatal damages to the aquatic systems.

Acid rain effects on forests can be devastating. It can kill trees by limiting the nutrients available to them, or poisoning them with toxic substances slowly released from the soil. Acid rainwater exchanges the nutrients in the soil needed by the plants (cations like Ca^{2+} and Mg^{2+}) or toxic ions (Al^{3+}, Cd^{2+}). The hydronium ions in acid rain can mobilize the latter in the soil and leach away the former, thus causing adverse effects on plant growth.

Acid seawater also has a corrosive effect on limestone or marble buildings or sculptures. Sulfuric acid present in rainwater reacts with $CaCO_3$ according to:

$$CaCO_3 + H_2SO_4 \rightleftharpoons CaSO_4 + H_2CO_3.$$

And, as we have learnt in previous sections, carbonic acid decomposes, producing $CO_{2g} + H_2O_l$ and calcium sulfate

($CaSO_4$), which is soluble in water. Therefore, the limestone (mostly $CaCO_3$) dissolves and crumbles.

Another important natural meteorological phenomenon that becomes harmful as a consequence of the presence of strong acids in the atmosphere are fogs, which consist of visible cloud water droplets or ice crystals suspended in the air at or near the Earth's surface. Depending on the cooling mechanism, there are a number of different kinds of fogs — radiation fog, ground fog, steam fog, ice fog, etc.

When acid pollutants are present in the air, the fog becomes acid, and because it is a type of a low-lying cloud its presence in the air greatly affects the breathing of all living beings on the planet. Acidic fog can be more hazardous to health than acid rains, as small droplets can be inhaled. A topical harmful effect of a fog is the so-called "killer fog" that hit London for several days in 1952. At least 12,000 people were killed, with around 150,000 hospitalization cases. Thousands of undocumented annual deaths also marked this fatal event.

CHAPTER 3

The Unique and Outstanding Properties of Water

Water is probably the most studied material on Earth, and the scientific study of its behavior and unique properties is ever growing. We are dealing with molecules of a small size and simple structure, yet water as a liquid exhibits an astonishing diversity of singular properties and capabilities, which seems to be preconceived to fit accurately the immense requirements of biological processes. Life depends on these unique properties of water. We will describe some of these properties and show how they result from the "simple" water structure and what vital role they play in maintaining life.

It should be noted that explanations of the complex behavior of water are not entirely satisfying and fully understood despite expanding research and knowledge. Some new theories appearing in the scientific literature are controversial and call for further work. Those theories are waiting for more experimental research and theoretical analysis, before being accepted by the scientific community.

We will refer briefly to some of the outstanding anomalous water properties on which life depends, before going into details and explaining how these properties originate from molecular structure and interactions.

- Water **conducts** heat more easily than any other liquid except mercury. This fact causes large bodies of liquid water like lakes and oceans to essentially have a uniform vertical temperature profile.

- The high **cohesion** forces between molecules (attractive forces that exist between molecules of the *same* type without losing their identity) are responsible for the great gap between the freezing and boiling points of water. Thus, water molecules exist in **liquid form** over an extended range of temperatures (from 0°C to 100°C). This range allows water molecules to exist as a liquid in most parts of our planet.

- Water is a universal excellent **solvent**. It is able to dissolve a large number of different chemical compounds due to its polarity and electrical properties (high dielectric constant), particularly polar and ionic compounds and salts. This characteristic enables water to carry solvent nutrients in living organisms.

- Water has a high **surface tension**. In other words, water is adhesive and elastic, and tends to aggregate in drops rather than spread out over a surface as a thin film. Adhesion causes water to stick to the sides of vertical structures despite gravity's downward pull. Water's high surface tension allows the formation of water droplets and waves, plants to move water (and dissolved nutrients) from their roots to their leaves, and the movement of blood through tiny vessels in the bodies of some animals.

- The large **heat capacity**, high **thermal conductivity** and high water content in living organisms contribute to thermal regulation and prevent local temperature fluctuations, thus allowing for a relatively easy control of body temperature.

- The high **latent heat of evaporation** provides resistance to dehydration and produces considerable evaporative cooling.

- In cells, water is in contact with the membrane, proteins and macromolecules that exhibit **hydrophilic** surfaces (showing a strong attraction to water). That determines their 3D structures and hence their biological functions.

- Water **self-ionizes** (loses the nucleus of one of its hydrogen atoms to become a hydroxide ion, OH^-) and allows for easy proton exchange between molecules, thus contributing to the richness of the ionic interactions in biology.

- At 4°C water **expands** when either heating or *cooling*. This density maximum together with the low ice density explains why the freezing of rivers, lakes and oceans takes place *from the top*, so permitting survival of life at the bottom. A layer of ice forms on top of the liquid water, creating an insulating barrier that protects animals and plant life below the surface from freezing.

- Water **density variations** are caused by temperature and salinity. Surface radiation, evaporation, precipitation, formation and melting of ice modify the density of surface water. The denser water will sink in the form of oceanic convections. Density-driven thermal convection provides oxygen into the depths (taken from the surface). The denser water will sink in the form of oceanic convection. As the water leaves the surface, its oxygen content is slowly used up for biological activity.

- The large heat **capacity** of the oceans and seas allows them to act as heat reservoirs, such that sea temperatures vary only a third as much as land temperatures and so help moderate climate.

- Water's high **surface tension** (the tendency of fluid surfaces to shrink into the minimum surface area possible) plus its expansion on freezing causes the erosion of rocks, thus providing soil for agriculture.

3.1. Heat Capacity and Resistance to Warming

A remarkable and important property of water is its high capacity to absorb heat without a significant increase in temperature. This property is called the heat capacity, or the specific heat capacity. Water can absorb a lot of heat before it begins to get hot. We can loosely say that water "resists" changing its temperature. You can observe the resistance of water to warming via an example of boiling water in a pot. When the pot is empty it warms up very fast. When we have water inside the pot, we may become impatient waiting for the water to become warmer and warmer until its boiling point. The ocean is rather cold during winter. When the weather becomes much warmer at the beginning of the summer, water remains pretty chilly at first.

On the beach on a hot day, the sand is too hot to walk on. Still the water temperature is cool enough to enjoy the touch of it. Both absorb the same amount of heat energy, but the temperature of the water is much less than that of the sand.

Let us say a few words about the apparently self-evident concepts of heat and temperature. Heat and temperature do not represent the same concepts. Heat is a kind of energy transfer as a consequence of a difference in temperature. Temperature is just a measure of molecular motion, i.e., translation, rotation and vibration.

Energy entering a substance may cause other molecular changes not leading to motion like molecular *structural changes*. This energy is "wasted" as far as temperature rise is concerned. We shall discuss this point when explaining the origin of the high heat capacity of water.

The specific heat of a substance is defined as the amount of heat that must be added to one gram of that substance to raise its temperature by 1°C. The specific heat of water is 1 calorie, which is needed to raise 1 gram of water by 1°C.

As a matter of fact, we have just defined what is known as a "small" calorie. To get familiar with this unit of energy (remember that heat is a energy transfer), we will say that a "large" Calorie (with capital C) you read on food package is 1000 "small" calories as defined above; so one "small" calorie is not much indeed. But consider the number of grams of water inside your body — about 70% — and consider how much energy is needed for raising the temperature of the whole body by 1°C (on average, every tissue and cell has its own story). The body receives heat not only through sun radiation and contact with warm air (via two different ways of heat transfer: Radiation and heat flow from a warmer substance to a colder one), heat is also produced in a living organism (insects not included) by the metabolism that runs chemical reactions with heat production. Our bodies maintain nearly a constant temperature when we are healthy. The process of energy provided by the metabolism will be analyzed in further sections. Water constitutes a means for heat regulation in the body.

When two bodies not in thermal equilibrium are placed next to each other, heat transfer will take place from the warmer body to the colder body, until thermal equilibrium is established. The amount of heat transfer, Q, is $Q = cm\,(T_2 - T_1)$, where c is the specific heat, m is the mass of the body, and T_2 and T_1 are the final and initial temperatures of the body, respectively.

Let us try to understand the origin of that high specific heat of water.

Water's high specific heat can be traced back to hydrogen bonding. Increasing temperature means increasing molecular motion: Translations, vibrations and rotations. The energy, in the form of heat you put into any liquid is usually used to increase molecular motion. However, when hydrogen bonds are present, some of the energy will go into breaking the HB, thus only a small part of the heat will be used to heat the liquid and increase molecular motion. The incoming heat energy is therefore employed in *breaking the hydrogen bonds* instead of increasing the motion and consequently the temperature will not increase much.

An extreme case is ice melting. All absorbed heat is used for breaking bonds and changing the solid phase into a liquid state. As long as there are two phases at equilibrium, there will be no temperature increase. We can say that the heat capacity at this state is infinity. The same happens with boiling water. Water boils until full evaporation takes place, keeping the temperature constant. All the incoming heat is used to break bonds and letting molecules leave the liquid phase completely.

Summary: When heat is transferred to water some of the energy is used in breaking hydrogen bonds. This energy is not available to increase the kinetic energy of the water molecules, so the temperature of water does not rise as much as it would in a normal liquid with weaker intermolecular forces. Therefore, water must absorb more heat energy to raise its temperature. Water can absorb large amounts of heat energy before it begins to get hot. Similarly, as water cools, it releases a great deal of heat energy. The high specific heat of water is responsible for the ocean's ability to act as a thermal reservoir, moderating the temperature changes from a hot day to a cold night, and from winter to summer.

Figure 3.1 illustrates this point well.

3.1.1. *Dramatic effects on life*

A large heat capacity means resistance to temperature change. It contributes to thermal regulation of our bodies.

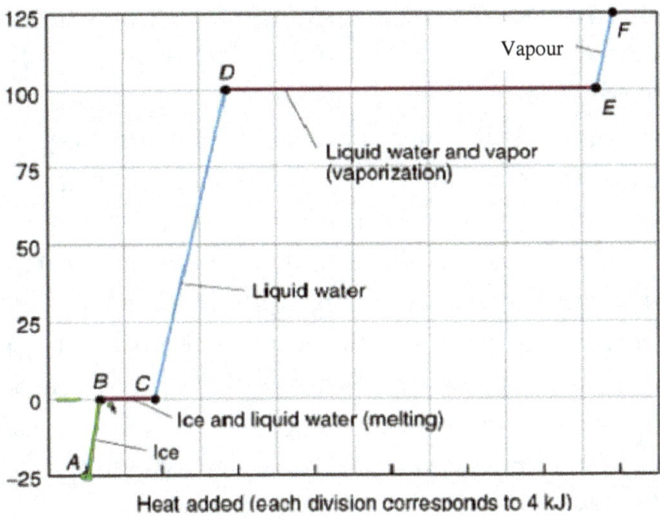

Fig. 3.1.   Variation of temperature as heat energy is added to water.

On a large scale, climate is moderated in coastal areas since solar heat energy in great amounts is absorbed by water during the day and temperature decreases only slowly at nights. Areas away from the coast experience much wider temperature variations. The deserts are the extreme case. Oceans are responsible for balancing the climate on Earth, permitting life to exist. Man moderates the temperature at his living place by building homes with insulating materials, proper clothes, heating and other cooling measures. For creatures living in water, pools, lakes and oceans, water is the natural moderator.

3.2. Boiling and Evaporation

Water exists in three phases: Solid (ice), liquid and the gas (see Chapter 1). Temperature and pressure are the external factors responsible for changing from one phase to another. Also the substances dissolved in water play their role. At the molecular level, the two factors are the energy added to or withdrawn from water molecules, and the nature of the interaction between water and the solved substance. Obviously the hydrogen bonding plays a decisive role here.

Transition from the liquid phase of the water to gas (vapor) is called evaporation. There are two types of vaporization: *Evaporation* and *boiling*. Evaporation is a "pseudo" phase transition from the liquid to vapor phase that occurs at the water surface without the appearance of bubbles at temperatures below the boiling temperature and at a rate that depends on the pressure.

How and why pressure? What is the role of temperature? What determines the boiling temperature? Did you know that water starts boiling at zero Celsius degrees under appropriate conditions? You know for sure that wet objects may dry up (evaporation!) under moderate temperature or even during chilly days, provided the air is dry enough. All this will be clarified in the next paragraphs.

Evaporation means escaping of water molecules from the surface. No boiling takes place if the temperature is under the boiling point. But some molecules may still occasionally enter back into the liquid phase, that is *condensation*, the opposite of evaporation (see Fig. 3.2).

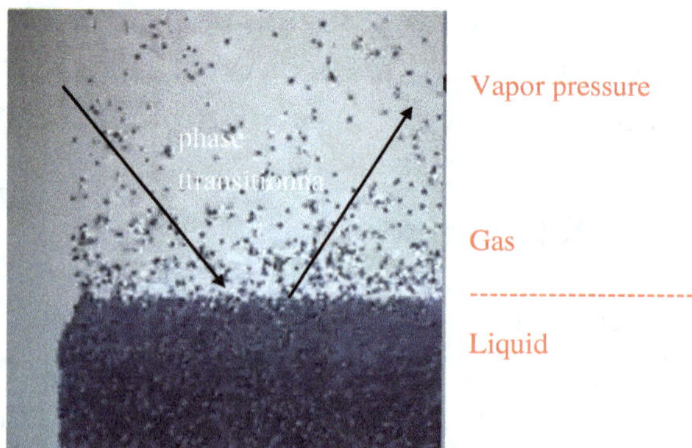

Fig. 3.2.   Evaporation (liquid → vapor) and condensation (vapor → liquid).

The rate at which water molecules enter back into the liquid depends on the concentration of water molecules above the surface. The gas phase is composed mainly of air molecules. *Water molecules are responsible for a small fraction of pressure in the gas phase.* If humidity in the air becomes larger, more water molecules are present in the gas phase and *evaporation is slowed down* (for given conditions, the amount of water molecules present during the gas phase is limited; see the saturated vapor water below).

We can isolate the effect of water gas pressure from the air pressure effect if we heat water in a *closed vessel*, which had been previously fully emptied of air (see Fig. 3.3).

The rate of evaporation in a closed container increases until there are as many molecules returning to the liquid as there are escaping. At this point the vapor is saturated, and the pressure of that vapor then is called the (saturated) vapor pressure. We say that liquid is at its equilibrium with its vapor.

The vapor pressure strongly depends on the temperature; it increases when the temperature increases. That energy should first break hydrogen bonds, thus enabling free rotation and translation motions. This gives water molecules the required energy to escape the surface.

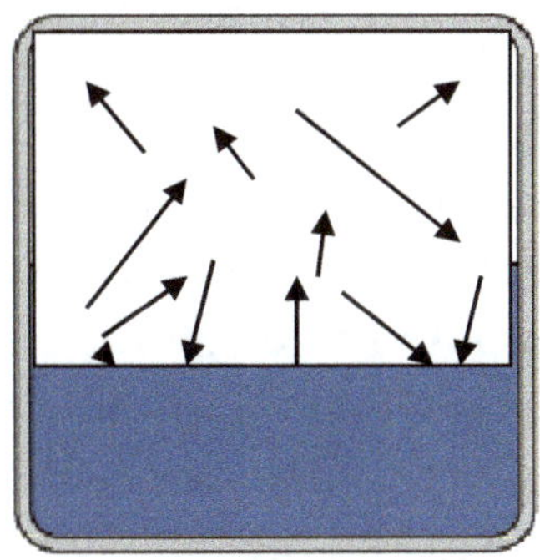

Fig. 3.3.   Evaporation in a closed container.

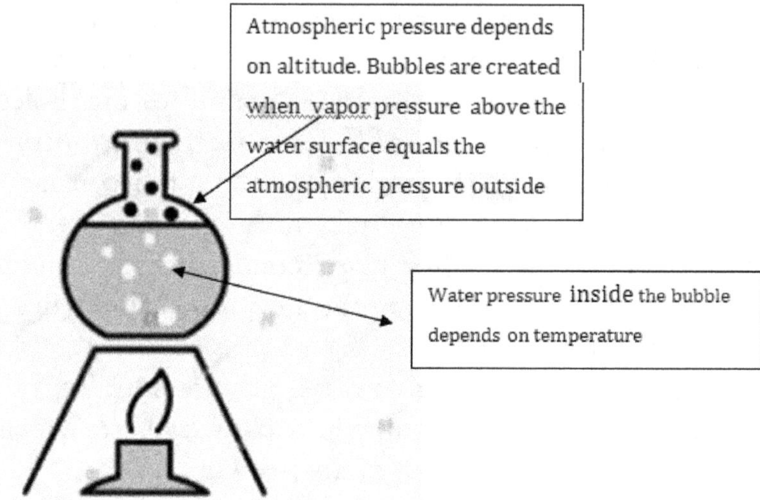

Fig. 3.4.   When vapor pressure above the surface increases and equals the atmospheric pressure, vapor bubbles are formed.

When the water surface is opened to the air, vapor pressure increases while the heat is transferred into the fluid. This keep occurring until the vapor pressure equals the atmospheric pressure. Then boiling starts, bubbles are formed and water molecules of water escape from the fluid!

The reason is that when vapor pressure equals the atmospheric pressure, bubbles can be formed (see Fig. 3.4) since they may overcome the outside pressure, which is approximately the atmospheric pressure.

Boiling is a phase transition from the liquid to gas phase that occurs while temperature remains fixed at the boiling temperature. Boiling, as opposed to evaporation, occurs below the surface. Boiling occurs when the equilibrium vapor pressure of the substance is greater than or equal to the environmental pressure. For this reason, the boiling point varies with the pressure of the environment.

It is well known that pressure decreases with increasing altitude (remember that the atmospheric pressure at any level in the atmosphere arises from the total weight of the air above a unit area at the point considered). It reduces to nearly zero (not exactly zero!) in outer space (see Fig. 3.5). The boiling point reduces to only 70°C at a height of 9 km. It is not very comfortable at the summit of Mount Everest, but if you wish to boil water there it will be quicker as compared to boling water on the Earth's level.

The heat of vaporization is the amount of energy necessary to transform one gram or one mole of a liquid into vapor at constant temperature and pressure. The heat of vaporization of water is 540 cal/g.

At the boiling point there is still a considerable amount of hydrogen-bonding ($\sim$75%) in water. Since all these bonds

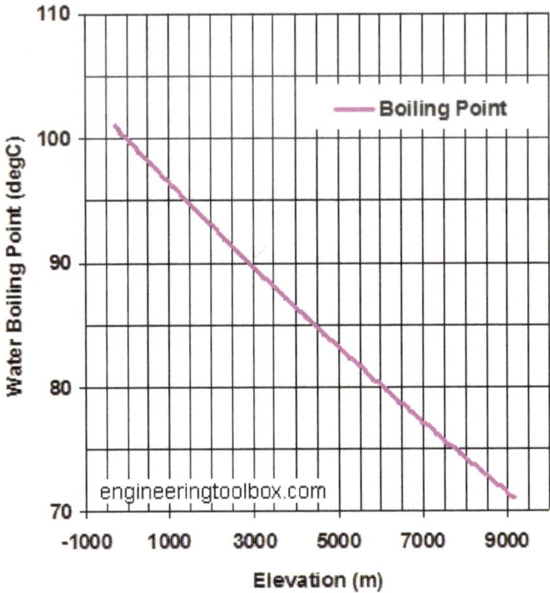

Fig. 3.5.   Variation of the water boiling point with height.

effectively need to be broken, there is a great deal of energy required to convert the liquid water to vapor. In the vapor phase, the water molecules are effectively separated.

Summary: The rate of evaporation, which is the number of water molecules that change from the liquid to the vapor phase each second, depends mainly on the temperature of the liquid water. The higher the temperature of the liquid water, the faster the rate of evaporation. Conversely, the rate of condensation, which is the number of water molecules that change from the vapor to liquid phase per second, depends mainly on the vapor pressure. The higher the vapor pressure the faster the rate of condensation. Condensation occurs when a water vapor molecule collides with a liquid water surface and binds to liquid water molecules. Evaporation occurs when a liquid water molecule attains sufficient energy to escape its chemical bonds with neighboring water molecules.

In a closed system, a dynamic equilibrium will be reached where the rate of evaporation equals the rate of condensation. When this occurs, the saturation vapor pressure can be measured and the air inside the closed system is saturated with water vapor. You may understand "saturation" to mean the maximum amount of water vapor that can exist in the air at a given temperature. It is very important to understand that the higher the air temperature, the higher the saturation vapor pressure. Saturation vapor pressure has strong temperature dependence. Figure 3.6 shows that a small increase in temperature results in a large increase in the saturation vapor pressure or capacity for water vapor in the air. In other words, warm air can hold more water vapor than cold air.

The relative humidity of the air is defined as the amount of water in the air relative to the maximum amount that the air can contain at that particular temperature.

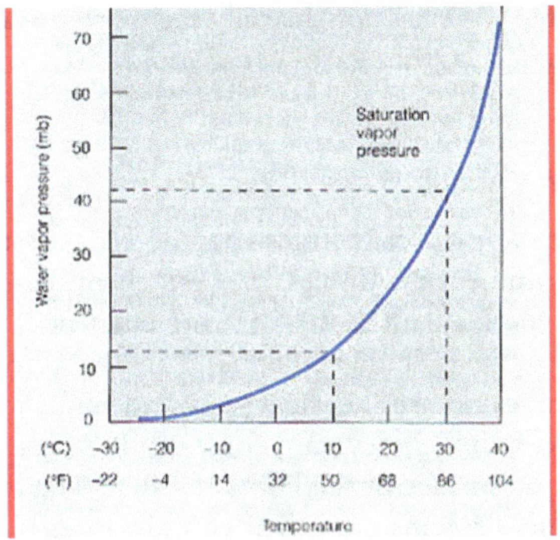

Fig. 3.6. Saturation vapor pressure increases with temperature. The pressure unit is millibar.

3.2.1. *Effects on climate*

Once water has evaporated, it becomes part of the gases that make up the atmosphere. The addition of water vapor to air reduces the density of air (the molar mass of water, 18 g/mole, is smaller than the molar mass of dry air, about 29 g/mole). A less dense air rises and cools. As air cools, vapor pressure decreases; in other words, the maximum amount of water vapor that the air can hold decreases. If the air rises high enough and cools sufficiently, it will not be able to hold all the water vapor it contains. When this happens, water vapor must condense back to liquid water. This is how clouds form. Clouds are composed of tiny droplets of liquid water (and possibly ice).

Water, due to phase changes from liquid to gas and vice versa, contributes a lot to climate. In addition, water vapor is a greenhouse gas that blocks energy radiation from leaving the surface of the Earth and helps to keep the planet warm enough to sustain life. Water is not the only factor responsible for the greenhouse effect. Other gases cause nearly 50% of the greenhouse effect.

3.2.2. *Cooling the body due to evaporation*

When we make physical efforts, our body heats up (even in a cold environment). Muscles convert chemical energy into mechanical energy with a 20–25% of efficiency. The excess energy is lost as heat. On the other hand, metabolic reactions are also exothermic and produce heat, which the body must remove in order to keep its temperature constant. Blood stream moves that heat to the surface of the body, the skin. There is a danger of overheating. Proteins need their comfort range for functioning. If the temperature becomes too high the proteins might be deformed and stop their vital functions. The large surface area of the skin allows part of the excess heat to be lost to the surrounding air. On the other hand, the hypothalamus in the brain is the body's thermostat. It stimulates sweat glands in the skin to produce fluid. Sweat drops appear then on the skin. The more heat that accumulates in the body, the greater the sweat on the skin. As the sweat evaporates from the skin, it removes heat (heat of vaporization) and cools the body. One should never forget that sweating means fluid leaking from the body. Therefore, one must drink enough water while doing physical activities. Note that when it is not too hot, sweat evaporates before accumulating on the skin. As more sweat evaporates from the body, the air close to it becomes more humid, making it harder for the sweat to evaporate.

Evaporation plays a critical role in preventing lakes and oceans from becoming too hot. Sunlight imparts a lot of energy to water and, consequently, raises its temperature. Without a mechanism to shed most of this heat, many (perhaps all) water-bodies would become too hot for most marine life.

Earth avoids this hellish fate by evaporating water. A portion of the liquid water is converted to water vapor, which carries away a lot of heat and cools the remaining water. To put this phenomenon in perspective, consider that tropical oceans lose an estimated 2.3 meters of water per year due to evaporation, which amounts to the absorption of about 1,000,000,000,000,000 (one quadrillion) calories per square kilometer (or about 2.6 quadrillion calories per square mile). In the case of lakes, evaporative cooling typically restricts the actual daytime temperature increase to only about 0.6°C per hour.

Water's high heat of evaporation also provides another mechanism for moderating global temperatures. Evaporative cooling in the tropics results in warm, moist air. Similar to ocean currents, wind transports that air to higher latitudes in the north and south, where it helps warm colder climates. This transfer of heat from hot to cold regions helps balance world temperatures. Hence, without evaporative cooling, planetary temperatures would be subject to far greater changes.

3.3. Cohesive and Adhesive Forces

Cohesion is the property of like molecules (of the same substance) to stick to each other due to mutual attraction. Adhesion is the property of different molecules or surfaces that enables them to cling to each other. For example, solids have high cohesive properties so they do not stick to the surfaces they come in contact. On the other hand, gases have weak cohesion. Water has both cohesive and adhesive properties. Water molecules stick to each other to form a sphere. This is the result of cohesive forces. When contained in a tube, the water molecules touching the surface of the container are at a higher level. This is due to the adhesive force between the water molecules and the molecules of the container.

The attraction responsible for cohesion originates from the electric structure of the molecule. Water is strongly cohesive since each molecule may make four hydrogen bonds to other water molecules in a tetrahedral configuration. This results in a relatively strong Coulomb force between molecules. In simple terms, the polarity (the state of which a molecule is oppositely charged on its poles) of water molecules allows them to be attracted to each other as explained in Chapter 1. In the case of a water molecule, the hydrogen atoms carry a relative positive charge in comparison to the oxygen atom. This charge polarization within the molecule allows it to align with adjacent molecules through strong intermolecular hydrogen-bonding, rendering the liquid cohesive. The high cohesion between water molecules is responsible for surface tension, which will be analyzed in the next section.

Water molecules carry permanent electric polarity due to its structure. Other materials may have a weaker attractive force, also electric in nature though their structure has no electric dipole. Their polarity is induced by the presence of neighbor molecules. Such forces are called van der Waals forces. They lead to a weak cohesion that operates by induced polarization in non-polar molecules (see Fig. 3.7).

3.3.1. *Van der Waals force*

When two neutral molecules have permanent electric dipoles, the tendency of such permanent dipoles to align with each other results in a net attractive force (dipole–dipole interaction). However, an induced attractive force can also occur between non-polar molecules, e.g., the noble gas argon or the organic liquid benzene. These forces of attraction between molecules account for condensation to the liquid state at sufficiently low temperatures. The nature of this attractive force

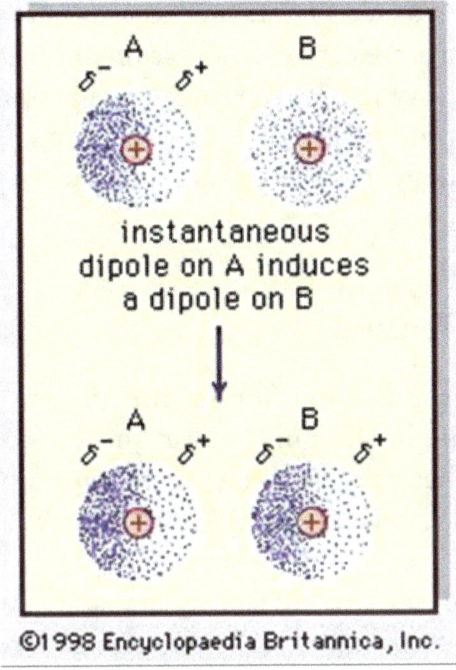

Fig. 3.7.   Van der Waals interactions.

in molecules — which requires quantum mechanics for its correct description — was first recognized by the German-born
(in Breslau, now Wroclaw, Poland) physicist Fritz London,
who in 1930 traced it to electron motion within molecules.
London pointed out that at any instant the center of negative
charge of the electrons and the center of positive charge of the
atomic nuclei would not be likely to coincide. Thus, the fluctu-
ation of electrons makes molecules time-varying dipoles, even
though the average of this instantaneous polarization over a
brief time interval may be zero. Such time-varying dipoles, or
instantaneous dipoles, cannot orient themselves into alignment
to account for the actual force of attraction, but they do induce
properly aligned polarization in adjacent molecules, resulting

in attractive forces (see Fig. 3.7). These specific interactions, or forces, arising from electron fluctuations in molecules (known as London forces, or dispersion forces) are present even between polar molecules and produce the net long-range attractive force, which is called the van der Waals force between the two neutral molecules.

3.4. Surface Tension

We show in Fig. 3.8 the different interactions experienced by a water molecule in the bulk and in the surface of the liquid. This difference between the forces experienced by a molecule at the surface and one in the bulk liquid gives rise to the liquid's surface tension.

The molecule at the surface is attracted by its neighbors below and on the surface. However, there are no attracting molecules above the surface. Molecules in the bulk of the liquid are fully surrounded by other molecules and apparently feel, on the average, no net force of attraction since each force exerted by a near molecule is supposed to be balanced by an opposite

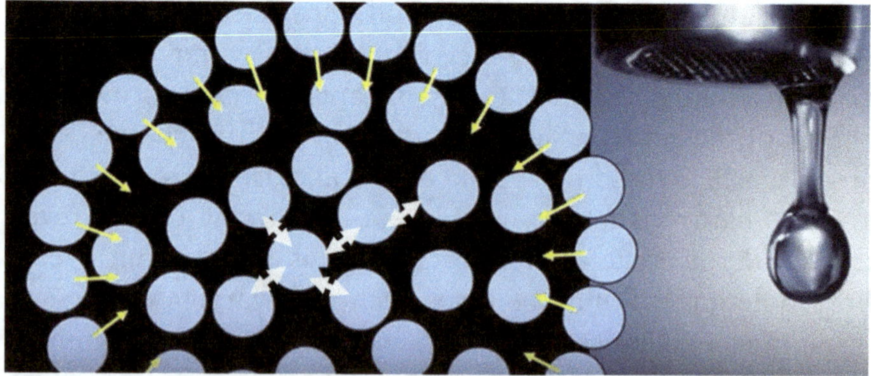

Fig. 3.8.   The molecules in the bulk experience a different interaction with their neighbors than those at the surface.

force. This static situation is an extreme idealization. In reality, molecules are constantly in thermal motion — one molecule moves a bit and another takes its place. Still there is a balance of forces on the average.

The molecules on the surface attract each other but are also pulled inward into the bulk of the liquid by the molecules on the inside of the liquid, and the overall effect is to minimize the surface area of a liquid. In other words, in order to increase the surface of a liquid we must fight (invest energy) against the surface tension forces.

3.4.1. *The angle of contact with a surface*

Let us consider water in contact with another substance through an interphase. If a water molecule is more strongly attracted by the rest of water molecules, then it will give rise to an increase of the curvature of the interface. This is what happens at the interphase between water and a hydrophobic ("water hating") surface such as a plastic mixing bowl or a windshield coated with oily material. A clean glass surface, by contrast, has —OH groups sticking out of it, which readily attach themselves to water molecules through hydrogen-bonding; this causes the water to spread out evenly over the surface, or to wet it. A liquid will wet a surface if the angle at which it makes contact with the surface is less than 90° (see Fig. 3.9). The value of this contact angle can be predicted from the properties of the liquid and solid separately.

The geometric shape that has the smallest ratio of surface area to volume is the sphere, so very small quantities of liquids tend to form spherical drops. As the drops get bigger, their weight deforms them into the typical tear shape.

If we want water to wet a surface that is not ordinarily wettable, we add a detergent to the water to reduce its surface

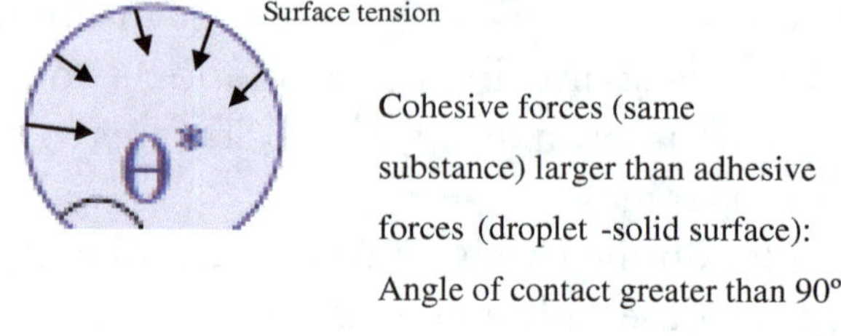

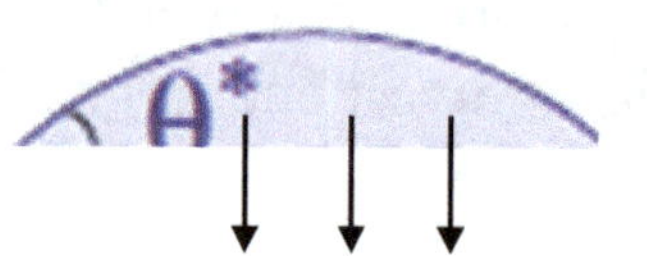

Fig. 3.9.   Liquid droplet in contact with a solid surface.

tension. A detergent is a special kind of molecule in which one end is attracted to water molecules but the other end is not, so these later ends stick out above the surface and repel each other (tending to increase the interphase surface), canceling out the surface tension forces due to the water molecules alone.

When a drop of a liquid is in contact with a solid surface, its behavior depends on the relative magnitudes of the surface tension forces (cohesive forces) and the attractive adhesive forces between the molecules of the liquid and the surface (see Fig. 3.10).

When we want to clean our clothes with fat stains, we will get the same result as with duck's plumage in the ponds if we were to use only water; water does not wet fat stains. However, if we add a soap or detergent, we get our clothes clean after rinsing. Soaps and detergents are substances with both hydrophilic and hydrophobic regions. The hydrophobic area can absorb the fat stain and at the same time water pulls the hydrophilic part towards the water bulk. Further rinsing dumps the dirty water into the sink.

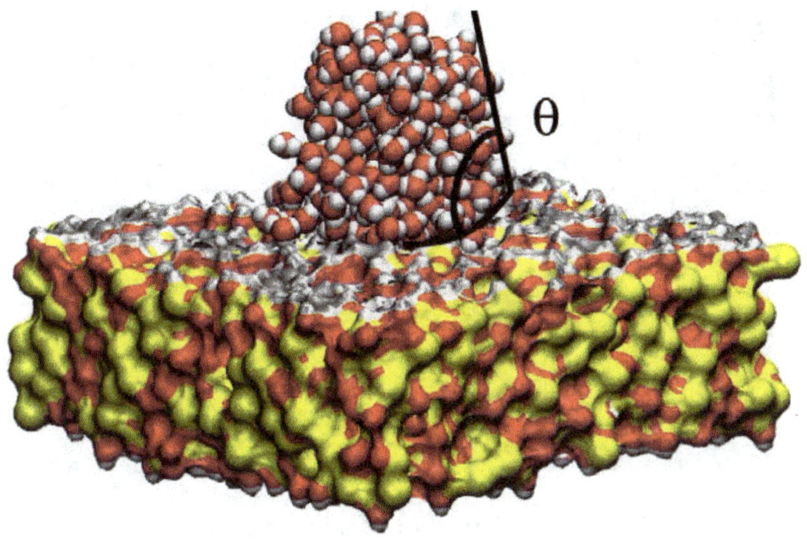

Fig. 3.10.   Water droplet resting on an inert surface. The contact angle is defined by the angle between the solid surface and the tangent to the surface liquid at the solid-liquid interphase.

Summary: The molecules on the surface behave in a way like a stretched elastic film that squeezes the liquid into the shape with the smallest possible surface area. The shape with the smallest surface-to-volume ratio is a sphere, so surface tension causes drops of water into as close as a spherical shape as possible (see Fig. 3.11).

Water striders use surface tension to walk on the surface of a pond. When the strider's leg (which is not attracted by water molecules because it is hydrophobic) pushes down on the water it deforms the surface. The surface tension of water pushes the water strider upwards in order to recover its original (smaller) surface shape, so keeping the smallest possible surface area to volume ratio. Thus, the water strider can stand on the water surface as long as its mass is small enough for the water can support it. Surface tension creates a membrane strong enough as to support the weight of a small insect (see Fig. 3.12).

Fig. 3.11.   Formation of a water droplet.

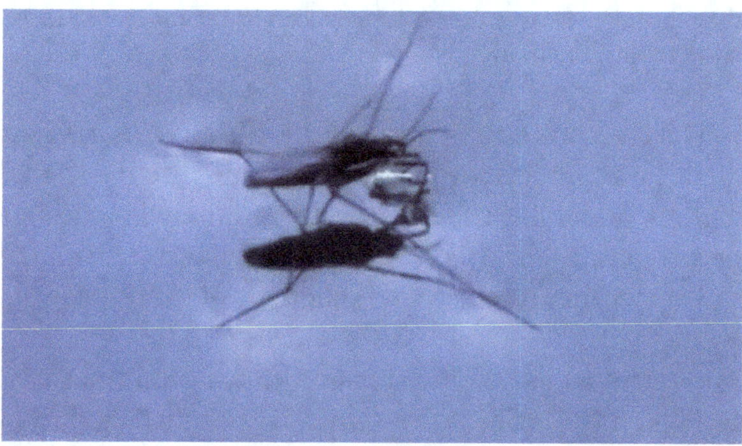

Fig. 3.12.   Surface tension creates on water a membrane strong enough to bear the weight of a small insect.

3.4.2. *Importance of capillarity to life*

How can water, despite the Earth's gravity, rise through gigantic trees against tenths of atmospheric pressure? This is due to the *adhesion* force; the attraction between the molecules of the liquid and those of the surrounding solid, which causes water to stick to the sides of vertical structures despite gravity's downward pull. Capillary action is the ability of a liquid to flow in narrow spaces without the assistance of, and in opposition to, external forces like gravity. If the diameter of the tube is sufficiently small, then more molecules are in contact with the inner surface of the tube; the combination of surface tension, mentioned above, and forces between the liquid molecules and container make the liquid lift up (see Fig. 3.13).

Capillarity is not restricted to glass tubes. Water climbs any dipolar material with charge on its surface. Water climbs the tiny fibers of paper towels and there is a constant pull from the roots to leaves of a plant through the thin vessels.

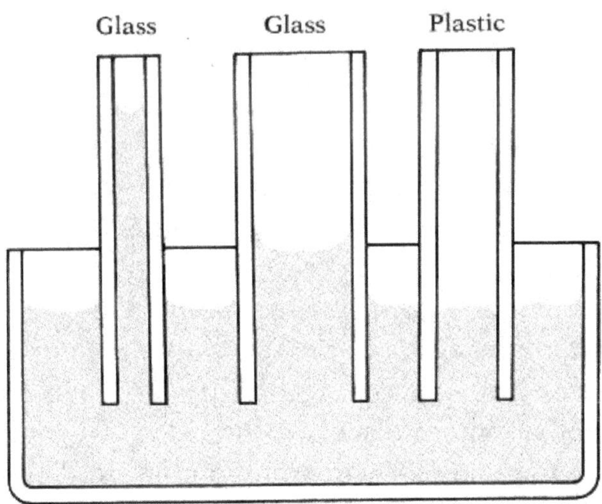

Fig. 3.13. Capillarity: Water rises higher in a glass tube of lower diameter because more water molecules are in contact with the glass in the narrower tube and can form hydrogen bonding with polar groups. Plastic has no charges on its surface and water molecules are not attracted to it. No adhesive forces act.

When water molecules reach the leaf, they may evaporate out into the air (transpiration). Water thus keeps moving in the plant vessels. Water plays an essential role in photosynthesis (see Appendix A). Life depends on that phenomenon — plants, and of course, consequently, our own lives.

The water's surface tension contributes to the formation of waves. As wind blows across the smooth water surface, friction between the air and water tends to stretch the surface, resulting in wrinkles. Surface tension acts on these wrinkles to restore the smooth surface, and this creates ripples (small waves). As the wind continues blowing, more and more big waves are generated.

3.5. Water Density Anomaly

Solid water (ice) does not sink into liquid water but rather floats upon it. As a rule, liquids contract when they get colder and reach a maximum density upon solidifying. It is different with water: When cooled, water contracts until it reaches the temperature 4°C. Then, it expands until reaching its freezing point at 0°C. At freezing temperature, ice is less dense than water. That means that the mass per unit volume becomes smaller. Volume increases at freezing temperature by about 9% under atmospheric pressure. If the melting point is lowered due to an increase in the pressure, the increase in volume upon freezing is even greater (for example, 16.8% at −20°C). This is due to the "cooling" of intermolecular vibrations, thus allowing the molecules to form steady hydrogen bonds with their neighbors and thereby gradually locking into positions that resemble the packing achieved when water freezes to ice.

The density maximum of water is a dramatic expression of the central role played by hydrogen-bonding in determining the properties of this liquid: As temperature decreases through

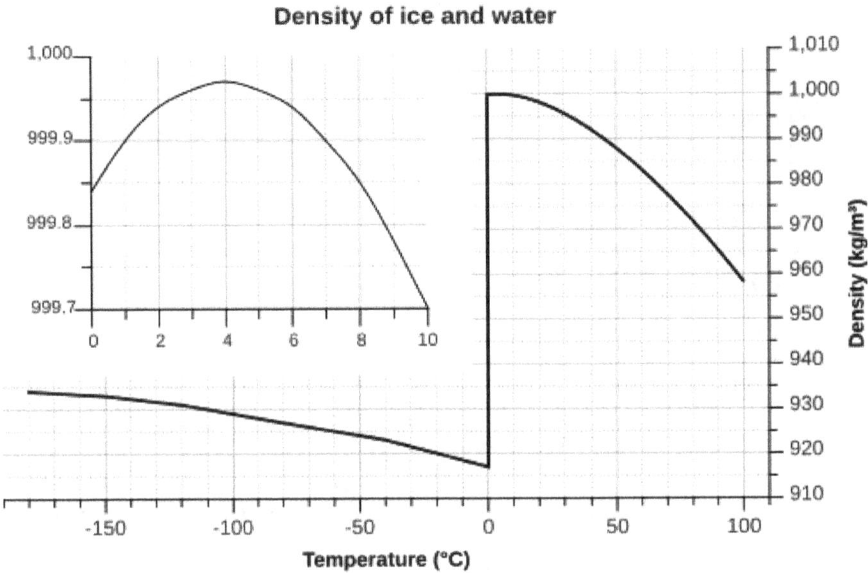

Fig. 3.14.   The variation of ice and water densities with temperature. There is a large decrease (about 9%) in density on freezing that represents a discontinuity at phase transition (liquid → solid).

the region of the maximum density, an increasingly organized network of hydrogen bonds expands the volume occupied by the liquid, overwhelming the normal tendency of the liquid to contract as it is cooled. As a consequence, the density of water decreases from 4°C (see Fig. 3.14).

When water becomes ice, its density starts to increase slightly as the temperature falls further below zero, though it never becomes as dense as liquid water. At 0°C, ice has a density of 0.9162 g/mL, and its density is 0.9257 g/mL at −100°C.

It is interesting to mention that the melting point of ice is 0°C (32°F, 273.15 K) at standard pressure. However, pure liquid water can be supercooled well below that temperature without freezing, if the liquid is not mechanically disturbed. It can remain in a liquid state down to approximately 231 K (−42°C). The

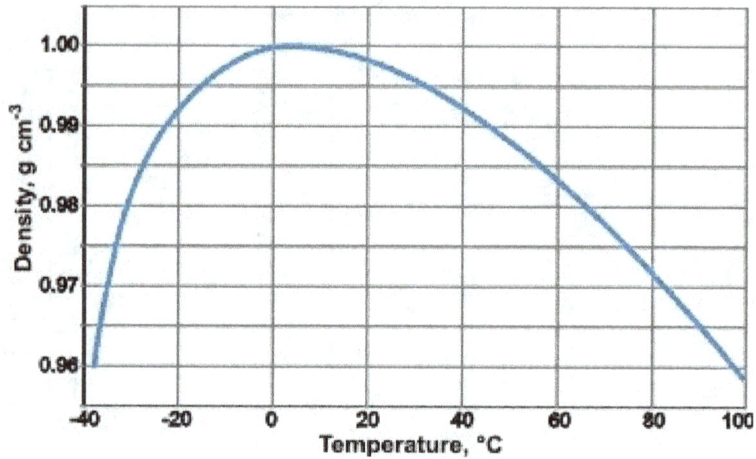

Fig. 3.15.   Variation of water and supercooled water with temperature.

variation of density with temperature for super cooled water can be seen in Fig. 3.15.

3.5.1. *The amazing story of freezing*

As the poem by Robert Frost reminds us, ice and fire are two ways of ending life:

Some say the world will end in fire,
Some say in ice.
From what I've tasted of desire
I hold with those who favor fire.
But if it had to perish twice,
I think I know enough of hate
To say that for destruction ice
Is also great
And would suffice.

However, the way ice is formed and its properties and structure under temperature change are vital to life, even though the iceberg caused the sinking of the *Titanic* in 1912, an event that inspired Frost to write this poem. As we shall see, life should be thankful to an iceberg. The key is the density of both ice and liquid water.

Water density anomaly has important environmental consequences since icebergs float in the ocean and lakes freezes from the top, so that aquatic plants and animals survive in the unfrozen liquid below.

If ice was heavier than water, all the ice that is formed in the northern regions would sink to the bottom of the Arctic ocean instead of forming an ice shelf that provides thermal insulation to the liquid below, and the production of ice would continue until the ocean is completely frozen form bottom up to the surface.

The anomalous expansion of water helps preserve aquatic life during very cold weather. When temperature falls, the top layer of water in a pond contracts, becomes denser and sinks to the bottom. A circulation is thus set up until the entire water in the pond reaches its maximum density of 4°C. If the temperature falls further, the top layer expands and remains on the top till it freezes. Thus even though the upper layer is frozen, the water near the bottom is 4°C and the fishes, etc., can survive.

Water at a temperature of 4°C will always accumulate at the bottom of freshwater lakes, irrespective of the temperature in the atmosphere. Since water and ice are poor conductors of heat (indeed they are good insulators) it is unlikely that sufficiently deep lakes will freeze completely, unless stirred by strong currents that mix cooler and warmer water and accelerate the cooling (see Fig. 3.16). This property therefore allows aquatic life in the lake to survive during the winter.

Fig. 3.16.   Freezing does not reach deep waters.

3.6. Ice: Structure and Phase Diagram

So ice is lighter than water. And it is "well known" that every material that is lighter than water floats on it. It is as simple as that. But not so fast! Why is that? What does floating exactly mean? There is a story about the revelation of that fact, an explanation based on classical physical laws and an explanation based on the molecular theory of water. We shall give some space to both views in the following section, including more about the structure of ice.

Scientists, educators, engineers, navigators, meteorologists and other professionals deal with ice. Some of these people contribute to the knowledge on which ice's physics is based. Successful applied ice research is based upon fundamental science. The first modern ice physics text was N. H. Fletcher's *The Chemical Physics of Ice* (1970). Fletcher's book is written in typical textbook format: It is reasonably brief and easy to understand. He touched on a few of the most important topics.

The most comprehensive book on ice physics to date is *Ice Physics* (1974), published by P. V. Hobbs, who considered almost all of the basic aspects of ice as understood at that time. Moreover, he described and compared several (sometimes opposing) viewpoints. This fundamental book is commonly known as the "Ice Bible" by specialists in the field. Since then, a significant amount of new experimental and theoretical work has appeared, dramatically changing our views on ice's physics.

3.6.1. *Ice Structure*

The crystal structure of ice is a result of hydrogen-bonding. When water freezes at atmospheric pressure, the interaction energy between molecules is reduced through the formation of a regular network.

Water normally freezes when it is cooled below 0°C, forming ice crystals. Ice crystals form more easily when they grow on existing ice crystals. A small piece of dust, or another type of impurity in the water, or even a scratch on the bottle are sometimes all it takes to get ice crystals growing. The process of starting off a crystal is called "nucleation".

Ice has many forms. In all of them, ice, like all solids, has a well-defined structure in which the central unit is a motif of a water molecule hydrogen-bonded to four water molecules, as shown schematically in Fig. 3.17. Each water molecule is surrounded by four neighboring water molecules. Two of them are hydrogen-bonded to the oxygen atom on the central water molecule, and each of the two hydrogen atoms is similarly bonded to another neighboring water molecule.

To "feel" the "realistic" 3D structure of the base unit, take a look at Fig. 3.18.

The ordinary ice we are familiar with that we also see in frozen lakes is called ice I_h (hexagonal ice; pronounced ice one h, also known as ice-phase-one). The whole crystal can be built

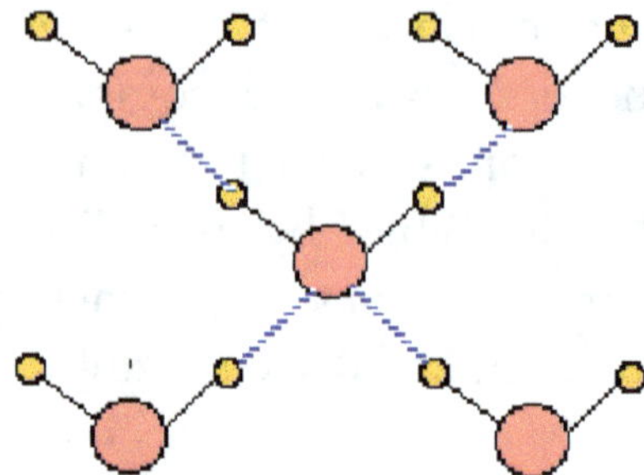

Fig. 3.17. The hydrogen bonds are represented by dashed lines in this 2D schematic diagram. In reality, the four bonds from each O atom point toward the four corners of a tetrahedron centered on the O atom. This basic assembly repeats itself in three dimesnions to build the ice.

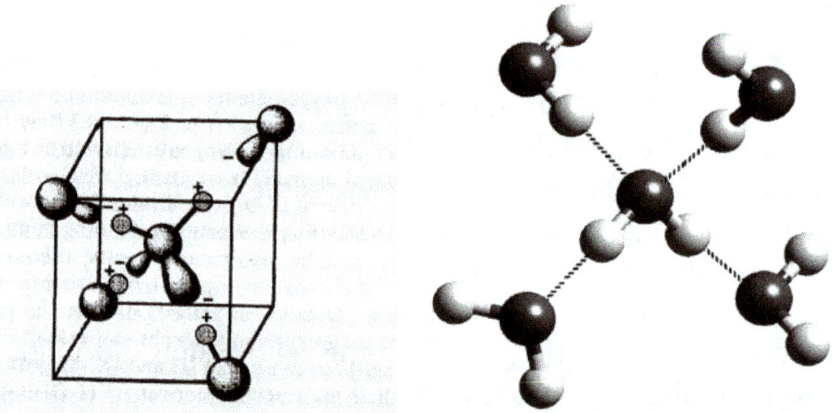

Fig. 3.18. Five water molecules showing the arrangement of hydrogen-bonding in crystalline ice. A four-coordinated water molecule showing the classic tetrahedral arrangement of the first-neighbor environment of a water molecule hydrogen-bonding to four neighbors. The central molecule contributes two hydrogen bonds to its two lower neighbors and "accepts" a hydrogen bond from each of its two upper neighbors. On the left, the same motif is presented using a chemical model.

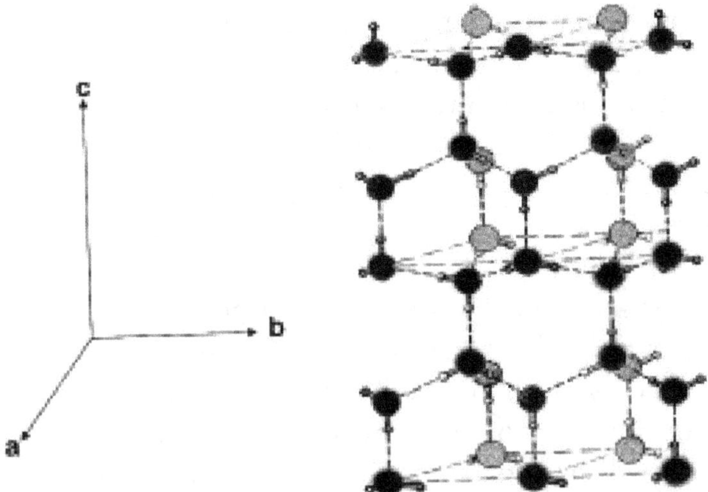

Fig. 3.19.   The structure of ordinary ice I_h. If you shrink yourself like Alice in *Alice in Wonderland* and sit on one of oxygen atoms, you should see four other oxygen atoms in the immediate vicinity at a distance of 2.76 Å. All four positioned at the corner of a regular tetrahedron. Reproduced with permission from John L. Finney.

from unit cells by the consecutive process of attaching one cell to another (see Fig. 3.19).

The hydrogen atoms in the crystal lattice lie very nearly along the hydrogen bonds, and in such a way that each water molecule is preserved. This means that each oxygen atom in the lattice has two protons adjacent to it, at about 101 pm along the 275 pm length of the bond. The crystal lattice allows for a substantial amount of disorder in the positions of the hydrogen atoms frozen into the structure as it cools to absolute zero. As a result, the crystal structure contains some residual entropy inherent to the lattice and determined by the number of possible configurations of hydrogen positions that can be formed, while still maintaining the requirement for each oxygen atom to have only two hydrogen atoms in closest proximity, and each hydrogen bond joining two oxygen atoms having only one hydrogen atom. This residual entropy is equal to $3.5\,\text{J mol}^{-1}\,\text{K}^{-1}$. A detailed analysis of Pauling's estimate of this value was carried out in Chapter 1.

The resulting structure is shown in Fig. 3.19.

The ice lattice is considerably less dense than the liquid water structure, to the degree that there is enough room in between molecules to contain other water molecules. This is the reason for the lower density of ice with far-fetched consequences to life, as has been mentioned in previous sections.

That is the only structure of ice found at the temperature and pressure we have here on Earth. At much higher pressures of tens or hundreds of atmospheres, we find many other structures. We will provide some further details on them in a later section. The reason for their existence is that under high pressures, the distance between atoms and the direction of the bonds change, affecting the whole crystalline structures. But first we will describe the so-called phase diagrams.

3.6.2. *Phase Diagram of Ice Water*

The thermodynamic properties of a system formed by a single substance (for example, water) are not independent. Thermodynamic variables are related by means of what we call an equation of state. The graphical representation of the equation of state for a one-component system is the so-called P-V-T surface. It is a rather complex representation (consideration of multi-component systems lead to considerably more and more complex representations). In order to introduce some simplification, the three projections (P-V, P-T and T-V) are often analyzed. Figure 1.8 in Chapter 1 shows the P-T projection of the water phase diagram.

The great physicist Josiah Willard Gibbs showed that the description of the equilibrium state of a non-reactive system requires the knowledge of a number of independent intensive variables F (also called degrees of freedom) given by the equation $F = C - P + 2$, where C and P stand for components and phases, respectively. Thus, for example, in the case of water, any point inside the liquid area ($C = 1, P = 1$) has two degrees of freedom, i.e., we need to fix the values of P and T to identify such a point (see Fig. 1.8 in Chapter 1). We can have water at $P = 1$ atm and $T = 25°C$ or at $P = 1$ atm and $T = 80°C$ (see the horizontal line $P = 1$ atm in Fig. 1.8 in Chapter 1).

In Chap. 1 we did a careful general analysis of the different regions in the phase diagram of water. Here we want to focus on ice water.

There are about 17 different forms of crystalline ice that we know experimentally, where the oxygen atoms are in fixed positions relative to each other but the hydrogen atoms may or may not be disordered. The hexaganol (I_h) form known as ice (see Fig. 3.20) *is the only one that is found naturally.*

The water phase diagram in Fig. 3.20 shows the behavior of ice forms of water at atmospheric pressure and much below. At the lower pressures in the diagram, there is one form of ice called I_h. At much higher pressures of hundred atmospheres, crystallization of ice leads to various structures denoted by Roman numbers as VI, X, etc. Ice X, for example, *has no water molecules*. It contains ions (or atoms) of oxygen and hydrogen.

The increase of pressure leads to the distortion of the tetrahedrons, which results in the increase of ice density (ices II, III, IX). Each form has its particular structure. It has nothing to do with life (unless life exists on stars where such pressures exist) but we show for completeness the conditions under which such forms of ice exist in the phase diagram of Fig. 3.20. At any pressure you can move towards higher temperatures between from $-200°C$ up to above $100°C$. You can move to higher pressures and find different forms of ice. We will not analyze those structures.

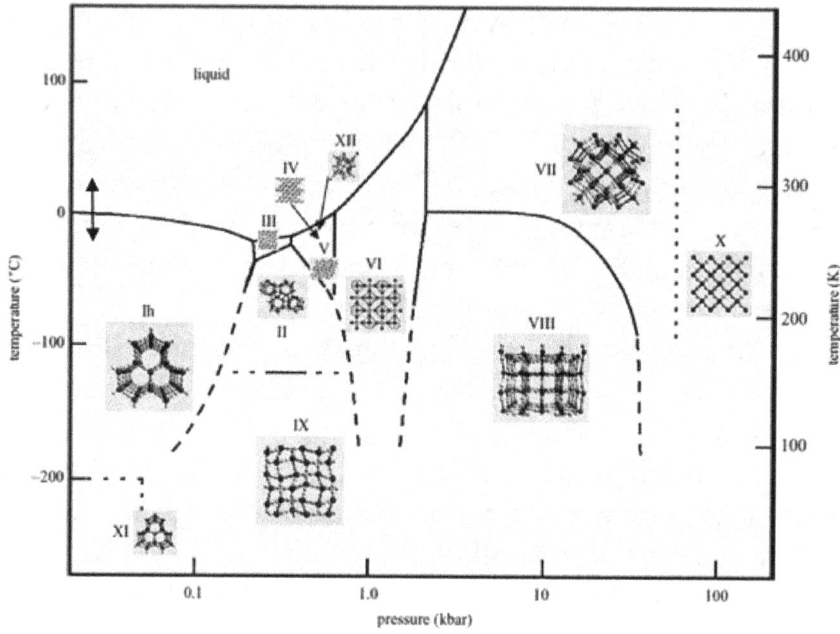

Fig. 3.20. Phase diagram at high pressures showing diverse structures of ice unseen under natural conditions (1 bar = 0.9869 atm).

3.6.3. *Snowflakes*

Snow crystals form when tiny supercooled cloud droplets (about 10 μm in diameter) freeze. These droplets are able to remain liquid at temperatures lower than $-18°$C. In order to freeze, a few molecules in the droplet need to get together by chance to form an arrangement similar to that in an ice lattice, before the droplet can freeze around this "nucleus". Experiments show that this "homogeneous" nucleation of cloud droplets only occurs at temperatures lower than $-35°$C. In warmer clouds an aerosol particle or "ice nucleus" must be present in (or in contact with) the droplet to act as a nucleus. Our understanding of what particles make efficient ice nuclei is poor. What we do know is that they are very rare compared to that cloud condensation nuclei on which liquid droplets form. Clays, desert dust and

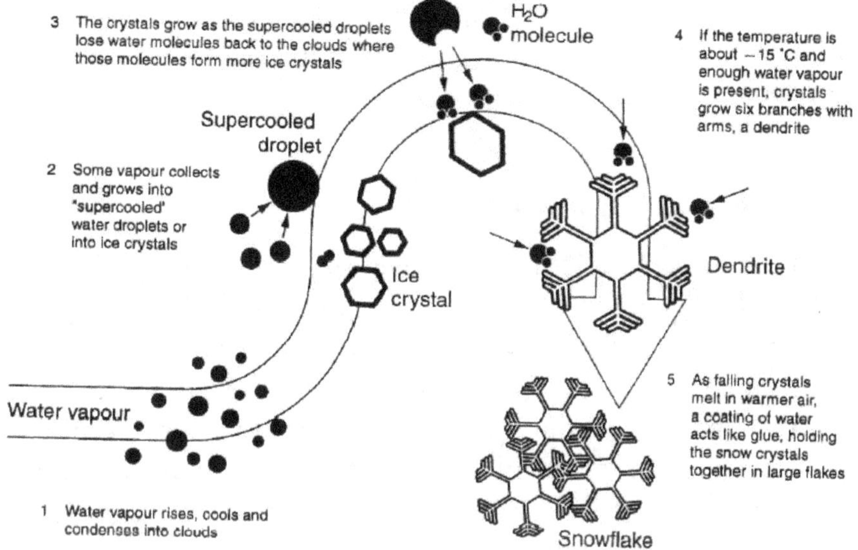

Fig. 3.21.   Plausible mechanism for snowflakes' formation.

Fig. 3.22.   Different snowflake shapes.

biological particles may be effective, although to what extent is unclear. Artificial nuclei are used in cloud seeding. The droplet then grows by the condensation of water vapor onto the ice surfaces.

Figure 3.21 provides details on the formation process of snowflakes and Fig. 3.22 shows some of the unique snowflakes' shapes. They all have a hexagonal crystalline structure but a unique geometry. The shapes of the flakes are determined by the atmospheric conditions experienced as they fall through the sky. Temperature and humidity are important factors causing variations in crystal growth.

3.6.4. *Freezing: Historical Note*

The fact that crystalline ice has lower density than liquid water seems to have been recognized for the first time by Galileo in 1612. He applied the basic principles of hydrostatics developed by Archimedes to the case of ice floating on water. In accordance with Archimedes' law of buoyancy, ice should have a lower density than liquid water. Galileo concluded: "*...in my judgment, the ice should rather be rarified water than condensed water; ... and water in freezing increases in volume, and the ice which is produced is lighter than the water on which it floats.*"

This conclusion, which is obvious nowadays, was questioned by the opponents of Galileo, relating the floating of ice to its shape. Galileo's idea was tested experimentally by a group of ten Italian "scientists" (Academy of Experiment, Florence, Italy, 1657–1667). The freezing of liquid water in various completely filled and tightly closed vessels caused the destruction of the latter, which clearly evidenced the expansion of liquid water upon freezing. By comparing the volumes occupied by water before and after freezing, the ratio of the densities of ice and of liquid water was found to be about 8/9. In October 1657, while performing experiments with water freezing in a glass bulb with a long, thin neck, Florentine "scientists" noticed a non-monotonous variation of the water level in the bulb's neck upon cooling: "*...water in the neck...steadily sank towards the bulb with a moderate speed until when it arrived at a certain level in continued down no farther, but stayed there for some time... Then it gradually began to rise again*". This was the first experimental observation of the density maximum of liquid water. The acceptance of this interesting finding by the scientific community took many years. Much skepticism, discussions and arguments on the pros and cons followed the progress of the theory that seemed so simple and self-evident in our days. The start and driving force of every revelation is what seems to be an unsolved mystery. That is true for many theories in biology, medicine, physics and chemistry. Scientific research becomes a "battle field".

3.6.5. *Archimedes and Pascal Laws*

Let us briefly mention two important principles that were formulated when analyzing certain "magical" behaviors of water, although they may apply to any fluid.

Pascal's law, named after Blaise Pascal (1623–1662), a French mathematician, physicist and religious philosopher and writer who was the founder of the modern theory of probability, states that the pressure at a point in a fluid at rest is the same in all directions. External pressure therefore, when applied on a fluid, is transmitted uniformly throughout its entire body. If we consider water in a container (see Fig. 3.23), the pressure depends only on the depth of the water. So pressure is the same in all parts of the container that are at the same depth. This explains the apparently paradoxical fact (hydrostatic paradox) that the water level for all the different shaped vessels in Fig. 3.24 is exactly the same.

At first sight, one could think that the greater the volume of the liquid is, the larger the pressure exerted on its base. What happens is that the force exerted by the liquid on the bottom of a recipient can be greater or smaller than the weight of the liquid column.

Let us now consider that there is water in the container in Fig. 3.23. Let us focus on a cylindrical volume element with

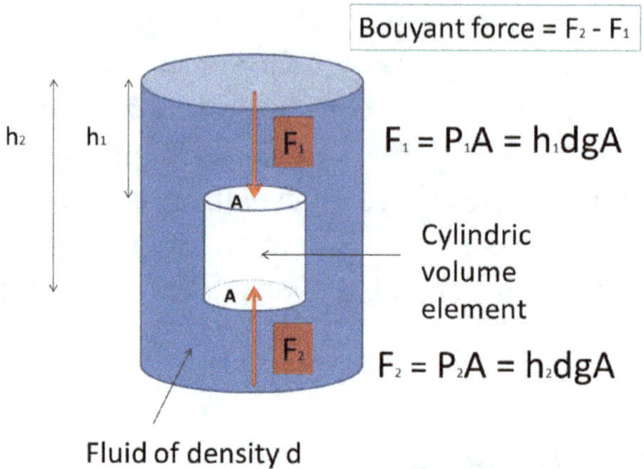

Fig. 3.23.   The pressure acting on an immersed body only depends on the depth.

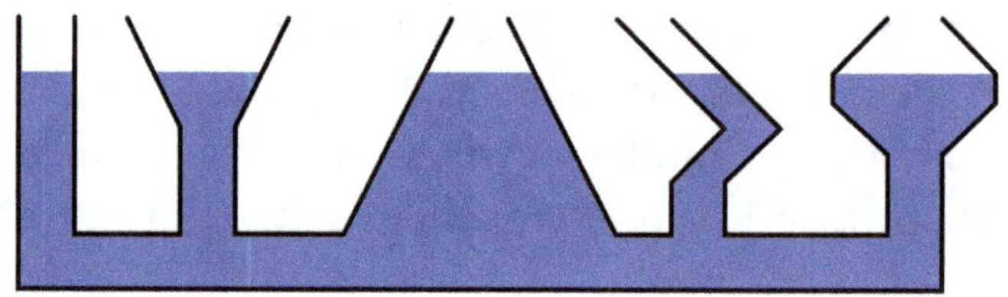

Fig. 3.24.   The water level is the same regardless of the shape of the vessel.

a mass of water $dA(h_2 - h_1)$ (d being the density of water). According to Pascal's principle, the different pressures at the top and bottom surfaces will give rise to a resultant force upwards (buoyant force $= F_2 - F_1$). If we had a cylindric body of any given material submerged in water and occupying an identical position of that of the cylindric volume element in Fig. 3.23, it should experience a similar upward force. Archimedes, a 3rd-century B.C.E. Greek mathematician and inventor, discovered this important result.

Legend has it that Archimedes discovered this law while sinking himself into a bathtub in a public bath. He then ran back

Archimedes' principle

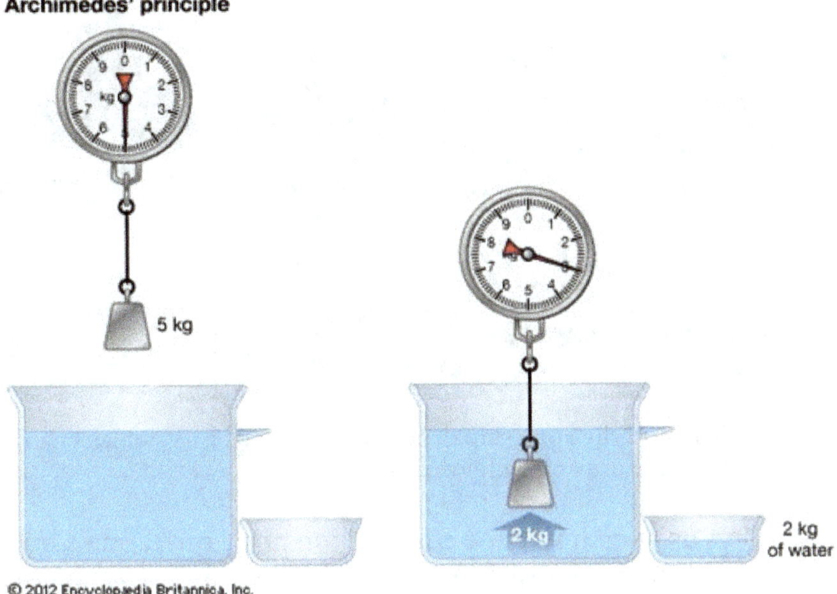

Fig. 3.25.   In the illustration, a 5 kg weight is fully immersed into the water, displacing 2 kg of water in the process. The amount of water displaced is not much, since the volume of the weight is small. Still the weight is reduced by 2 kg. Not enough for the weight to float.

home naked through the streets of Syracuse shouting "Eureka!" ("I have found it!"). The legend continues that he was able to apply this principle to determine whether a crown for King Hiero II was made of pure gold or had been adulterated with some cheaper metal (see Fig. 3.25).

One can conclude that if the density of the immersed object is greater than that of water, the object sinks. If the object's density is less than that of water, the object, as in the case of ice, floats. A rather simple mathematical relation that can be obtained in a straightforward manner is:

$$\frac{density\ of\ object}{density\ of\ fluid} = \frac{weight\ of\ object}{weight\ of\ displaced\ fluid}.$$

Fig. 3.26.   The waterline of a ship depends on water density, which changes with water temperature and water density.

This allows us to estimate the equilibrium position of one's body inside a fluid. For example, we can determine the waterline of a ship (see Fig. 3.26).

The depth of the immersing object depends on the ratio of the two densities. The density of ice is only slightly lower than that of water, so 90% of the iceberg's volume is below the water surface and can cause devastating damages (remember the *Titanic* and its 1514 deaths!).

Before ending, let us recall that Pascal's principle is the basis for one of the most important machines ever developed — the hydraulic press. The external pressure applied on a fluid according to Pascal's law is transmitted uniformly throughout its entire body.

Pressure is equal to the force divided by the area on which it acts. According to Pascal's principle, a pressure exerted on a piston in a hydraulic system produces an equal increase in pressure on another piston in the system. If the second piston has an area 10 times than that of the first, the force on the second piston is 10 times greater, though the pressure is the same as that on the first piston. This effect is exemplified by the hydraulic

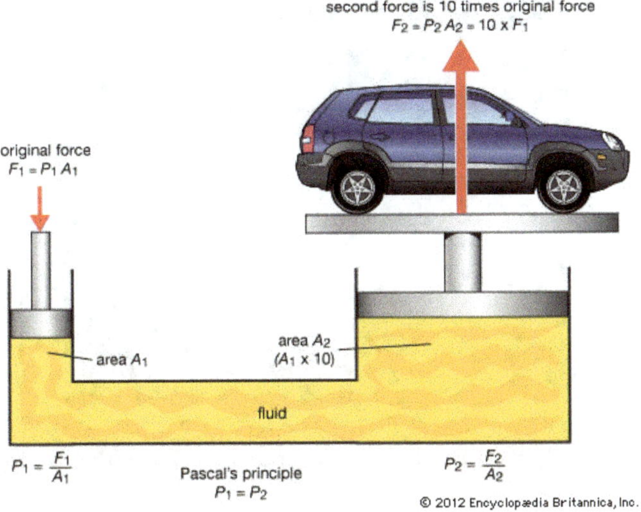

Fig. 3.27.   An illustration of Pascal's principle at work in a hydraulic press. According to Pascal's principle, the original pressure ($P1$) exerted on the small piston ($A1$) will produce an equal pressure ($P2$) on the large piston ($A2$). However, because $A2$ has 10 times the area of $A1$, it will produce a force ($F2$) that is 10 times greater than the original force ($F1$). Through Pascal's principle, a relatively small force exerted on a hydraulic press can be magnified to the point where it will lift a car.

press (see Fig. 3.27), based on Pascal's principle, which is used in applications such as hydraulic brakes.

3.7. Viscosity

Viscosity is the measure of fluid's resistance to flowing, or in other words, a measure of how sticky it is (see Fig. 3.28). Viscosity varies dramatically from one liquid to another; water has one of the lowest viscosities of all liquids. For example, the viscosity of tar is approximately 30 million times higher than that of water. Substances with high viscosities like honey or molasses (each about 10,000 times more viscous than water) pour very slowly. Viscosity also depends on temperature, e.g., engine oil is

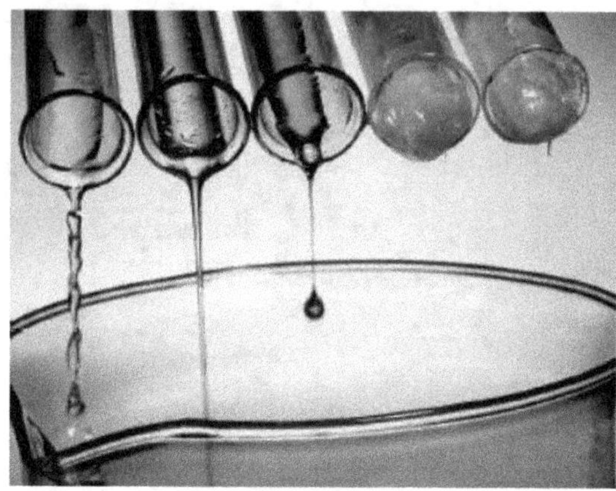

Fig. 3.28.   Effect of viscosity in pouring different substances.

much less viscous at high temperatures than it is in a cold engine in the middle of winter.

Viscosity is a property arising from the collisions between neighboring particles in a fluid that are moving at different velocities. When the fluid is forced through a tube, the particles that comprise the fluid generally move faster near the tube's axis and more slowly near its walls. Therefore some stress (such as a pressure difference between the two ends of the tube) is needed to overcome the friction between particle layers and keep the fluid moving. For the same velocity pattern, the stress required is proportional to the fluid's viscosity.

Water's viscosity is precisely in the range needed for life. If the viscosity of water were too high (like that of molasses), then it would impede blood circulation. Even a slightly higher viscosity would significantly increase the amount of energy needed to pump blood. Similar problems would occur within cells because a higher viscosity would hinder the movement of the large biological molecules and organelles required for cellular processes. A higher viscosity also creates more resistance for aquatic life to fight against while swimming, thus costing more energy.

A modest viscosity helps a cell maintain its shape, protecting it against sheering forces. If water's already low viscosity were even lower, external shocks and stresses would cause greater disruption to the cell. So for a liquid to serve as an effective medium for cells, its viscosity must be just right — neither too high nor too low. Water's viscosity fits the bill perfectly.

Viscosity of Ice

It turns out that the viscosity of solid ice must also be within a critical range. As it is, snow and ice near the poles accumulate into glaciers pliable enough to flow downward over time. These glaciers have played a key role in carving valleys and crushing rock into powder (which releases vital minerals into the environment). If ice's viscosity were too low, then ice would flow easily over land rather than transforming it. If ice's viscosity were too high, water would accumulate at the poles as ice, depriving the world of much-needed liquid water. Even a modest increase in ice's viscosity would increase glacier size (glaciers currently hold about 1.7% of the Earth's water), thus reducing the planet's amount of liquid water. Rain levels would drop and the global temperature would change more quickly. Fortunately for life on Earth, the viscosity of ice is fine-tuned to meet the needs of the planet and its inhabitants.

3.8. Diffusion

Diffusion is the reverse side of viscosity. It represents how quickly molecules spread out in a solvent. Imagine placing a drop of black ink in a glass of water. Initially, the ink drop starts in one place and then it spreads out gradually until it is distributed uniformly throughout the water. A substance's rate of diffusion is inversely proportional to its viscosity, which means water has a remarkably high rate of diffusion.

Diffusion is particularly critical in cells. For example, diffusion typically spreads oxygen throughout the cell interior within as little as a hundredth of a second, thus making it available for cellular processes. The same process makes the cell's importation of vital nutrients and the elimination of waste material efficient. This means simple unicellular life can exist by diffusion alone without requiring a circulatory system.

A complete discussion of the fine-tuned aspects of water's diffusion rate can be found in Michael Denton's *Nature's Destiny*. Some of his observations are summarized here. First, without water's high diffusion rate, capillaries in animals would be unable to deliver oxygen and nutrients to cells effectively. Second, blood (water containing red blood cells and other components) is a non-Newtonian fluid, meaning that its viscosity decreases with pressure. So, when the heart beats faster, the blood's viscosity drops, making it easier for the heart to pump. This behavior is critical for an efficient circulatory system in mammals.

Viscosity is a measure of how thick (viscous) and sticky a liquid is. Viscosity then reduces the ability of a liquid to flow. Any liquid that can flow readily (such as water) will have a low viscosity. Liquids with a high viscosity (such as molasses and motor oil) will flow more slowly and with greater difficulty.

Viscosity can be measured by timing how long it takes for the liquid to flow through a capillary tube or how long it takes for a steel ball to fall through the liquid. At a molecular level, viscosity is a function of the attractive forces of the molecules of the liquid, and to a lesser extent, of the presence of structural components (such as long-chain molecules) that can become entangled (a form of steric hindrance). Temperature also greatly affects viscosity — as temperature increases, viscosity decreases. This is because higher levels of kinetic energy are more able to overcome the intermolecular attractive forces. Those properties are critical to regulating chemical traffic into and outside the living cell without which life cannot exist.

Diffusion represents how quickly molecules spread out in a solvent from areas of high concentration to areas of low concentration. It occurs via random motion. Net diffusion stops when there is a uniform distribution of particles, and equilibrium is then reached. Molecules continue to move but no net change appears (on average) in concentration.

3.9. Osmosis

The osmotic phenomena are very important in all living systems.

Let us explain what this phenomenon is by means of the experiment showed in Fig. 3.29.

The U-tube in Fig. 3.29 is divided by a permeable membrane. Initially (see inset in the middle) pure water (thin points) is in one arm and sugar solution (bold balls) in the other. When the

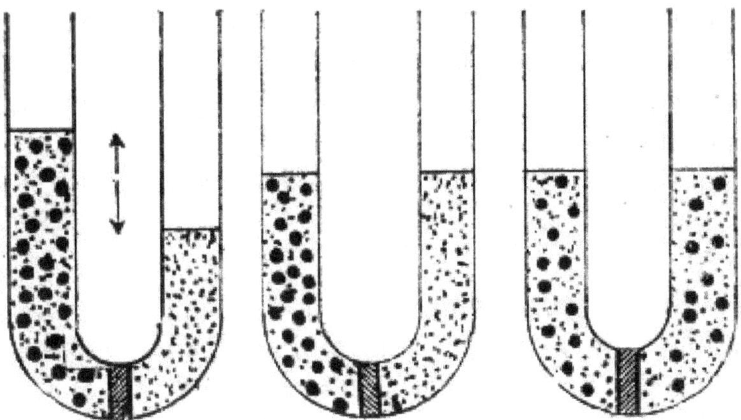

Fig. 3.29. Experiment showing the effect of osmosis.

membrane is permeable for both sugar and water molecules, an equal concentration of sugar in both arms is observed after some time. When the membrane is permeable to water molecules only (left inset), water penetrates the solution, sugar concentration reduces, and the total amount of solution increases until the height of the sugar solution prevents further level rising.

The solution level in the left inset rises until the hydrostatic pressure of the sugar solution prevents further penetration of water through the membrane. The osmotic pressure corresponds to the hydrostatic pressure exerted by the sugar solution at equilibrium. It is a measure of the tendency of water to move by osmosis into the solution. The more concentrated the solution is, the greater the tendency of water to move into it, hence the higher the osmotic pressure of the solution.

Let us suppose you were preparing for surgery: What would happen if you were injected with water instead of anesthetic? If plain water were injected into a vein, it would dilute the contents of the plasma (mainly electrolytes like sodium and chloride, and proteins) in that local area. Through osmosis, a portion of water would fairly rapidly be drawn into nearby cells, in particular, red blood cells. Many of these cells would expand in volume rapidly to the point that their membranes would rupture, spilling their contents into the circulation.

For a very small amount of water there would be no lasting harm. But as the amount increases, more red cells would be damaged.

However, if water with solution in roughly the same concentration as the plasma was introduced, then it would simply become part of one's circulating blood volume without causing any damage. The process described occurs when small volumes of saline are usually used to flush medication given through our intravenous line.

3.10. Cell Membranes: Diffusion and Osmosis in Living Cell

Coming back to cell life, both diffusion and osmosis are essential here since these are the mechanisms by which the transport of water and other molecules is made through the cell membrane into and out of the cell. That includes entering of oxygen into the cell and the expelling of carbon dioxide out of it, enabling life through breathing. We want to tell the reader a few things about the cell membrane structure and its role in letting essential molecules penetrate and exit the cell through diffusion and osmosis.

The cell membrane separates the interior of all cells from the outside environment, controls the movement of substances in and out of cells and protects the cell from its surroundings. It determines what can enter and leave the cell based on semi-permeability. Some molecules can enter the cell, others cannot. The cell membrane is selectively permeable to ions and organic molecules. Otherwise humans could not survive. Transport of water molecules through the cell membrane is certainly one of the basic subjects in cell biophysics.

That property of selectivity is due to the special structure of the cell membrane. We are going to describe the building blocks of the membrane, which are called phospholipids. What is behind that complex name is described in the next section.

3.10.1. *Lipids, Fatty Acids and Phospholipids*

We are going to describe them and show a little bit of their importance to cell structure and their "love–hate relations" with water.

"Lipids" is a term indicating a large group of organic compounds, including fats, oils and hormones (see Fig. 3.30).

$$
\begin{array}{c}
\mathrm{HO} \quad\quad \mathrm{O} \\
\diagdown \quad \diagup\!\!/ \\
\mathrm{C} \\
| \\
\mathrm{H}-\mathrm{C}-\mathrm{H} \\
| \\
\mathrm{H}-\mathrm{C}-\mathrm{H} \\
| \\
\mathrm{H}-\mathrm{C}-\mathrm{H} \\
| \\
\mathrm{H}-\mathrm{C}-\mathrm{H} \\
| \\
\mathrm{H}-\mathrm{C}-\mathrm{H} \\
| \\
\mathrm{H}
\end{array}
$$

Fig. 3.30.   Chemical structure of a lipid.

The common property of lipids is that they do not interact appreciably with water, contrarily to most of the molecular components of living organisms such as proteins, nucleic acids and carbohydrates, which are hydrophilic ("water loving") and soluble in water. In lipids that are part of the cell membrane, as we shall see, part of their structure is hydrophilic and another part, usually a larger section, is hydrophobic. They are called "amphipathic lipids" and exhibit a unique behavior in water — they spontaneously form ordered molecular aggregates with their hydrophilic ends on the outside, in contact with the water, and their hydrophobic parts on the inside, shielded from the water. This property makes them the basis for the cellular and organelle membranes.

The lipids that take part in a membrane structure are called phospholipids, since they contain a phosphate group at the end of their chain. It is worthwhile to get familiar with its structure: Phospholipids consist of three main components

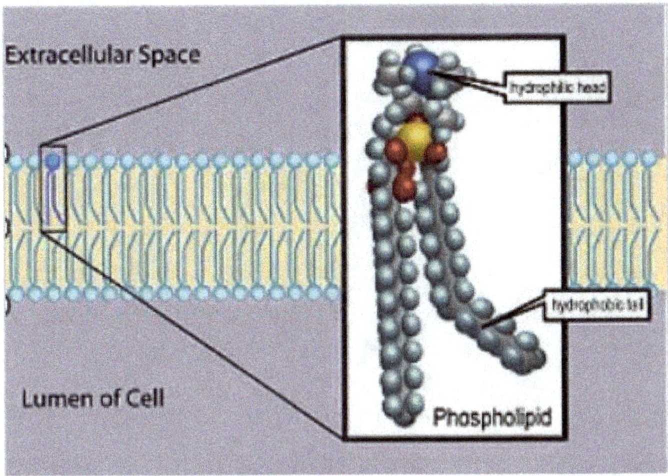

Fig. 3.31.   Schematic representation (left) and chemical structure (right) of a lipid.

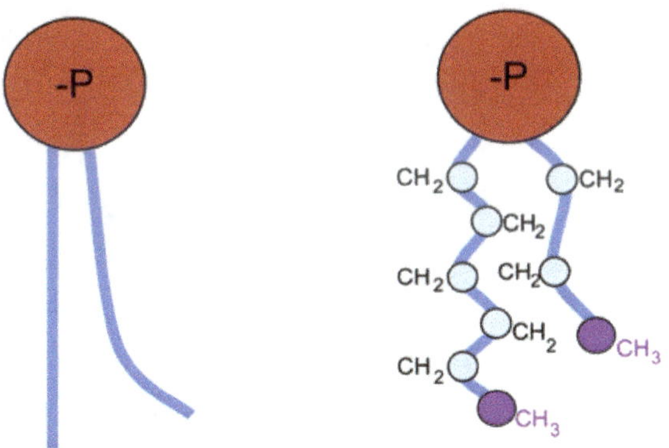

Fig. 3.32.   Schematic representation (left) and chemical structure (right) of a lipid.

(see Figs. 3.31 to 3.33):

- Phosphate head,
- Backbone (Glycol), which is between the two tails (unseen in the picture),
- Tails that are lipids.

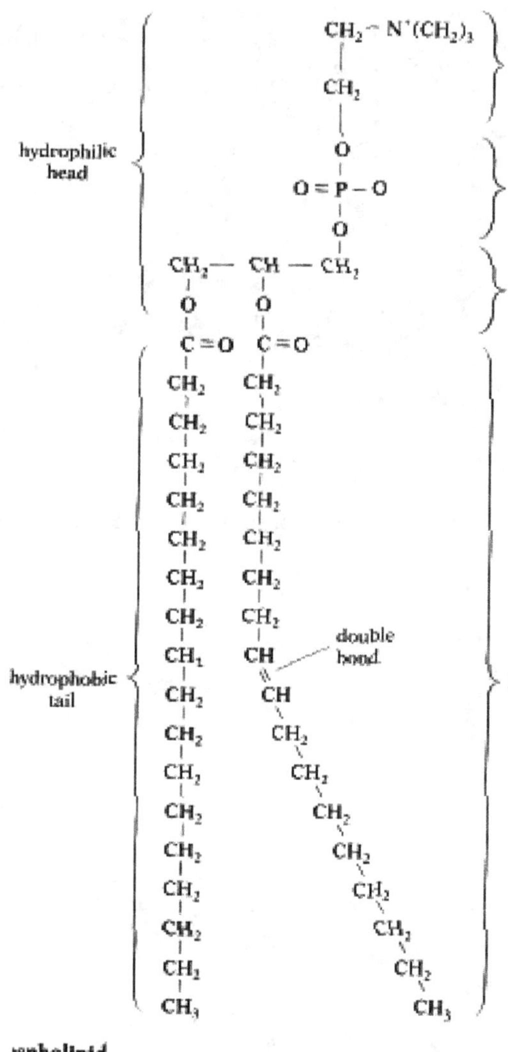

Fig. 3.33. Detailed chemical structure of a phospholipid. It consists of a polar head (a positively charged amine, a negatively charged phosphate, and an uncharged glycerol) joined to two hydrophobic fatty acid chains. The "kink" in the right tail is created by a double carbon bond. Because this tail is *unsaturated*, the phospholipid will be less tightly packed in the membrane and hence will be more mobile.

The head is polar hydrophilic, or "water loving". The phosphate head group is starving for water. The acid tails are composed of long carbon chains and therefore are hydrophobic, or "water fearing". These two fatty acids try to avoid water as far as they can. If one puts those two-sided molecules in the water, the hydrophilic heads will try to get closer to water as far as they can. The hydrophobic, on the contrary, will try to adopt a position far away from water. As a result, the phosphate groups cluster together while the tails try to shield themselves away from water. How can they accomplish this?

Figure 3.34 provides a detailed chemical structure of a phospholipid and Fig. 3.35 shows the effects of water on the phospholipid structure.

3.10.2. *Structure of the Cell Membrane and Their Selectivity*

A unique structure is shrewdly formed. *Two layers* (*bilayer*) of phospholipids are created, the polar (hydrophilic) heads of each

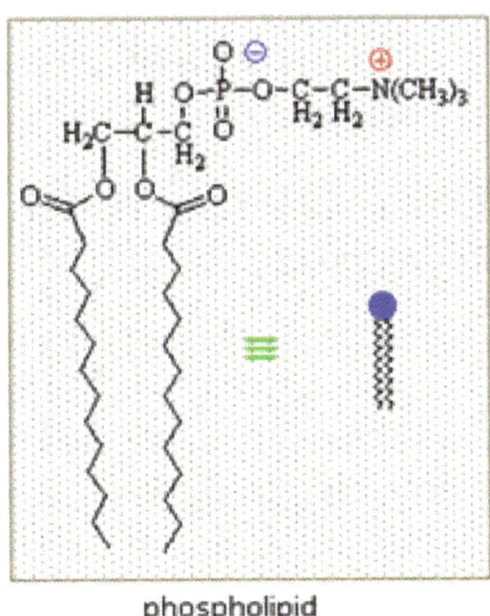

Fig. 3.34.   Chemical structure of a phospholipid.

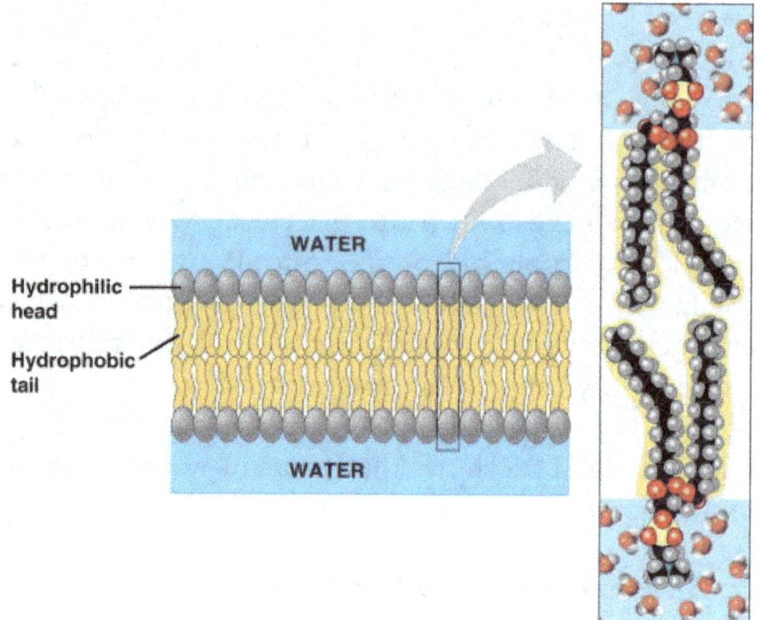

Fig. 3.35.   Phospholipid in water.

are exposed to the water and form the two external surfaces of the membrane, while the hydrocarbon tail chains are buried into the interior of the cell, hidden from the surrounding water. The structure is very stable.

3.10.3. *Fatty Acids, Metabolism and Metabolism in Aqueous Environment*

Water has a role in fat metabolism. Fats are important as energy storage molecules in living organisms. Most of the fat in the human diet consists of three fatty acids linked to glycerol. The chemical picture is shown in Fig. 3.36.

The key issue in the digestion and absorption of fats is related to solubility: Lipids are essentially hydrophobic, and thus are poorly soluble in the aqueous environment of the digestive tract. Fat breaks down in the digestion process again into

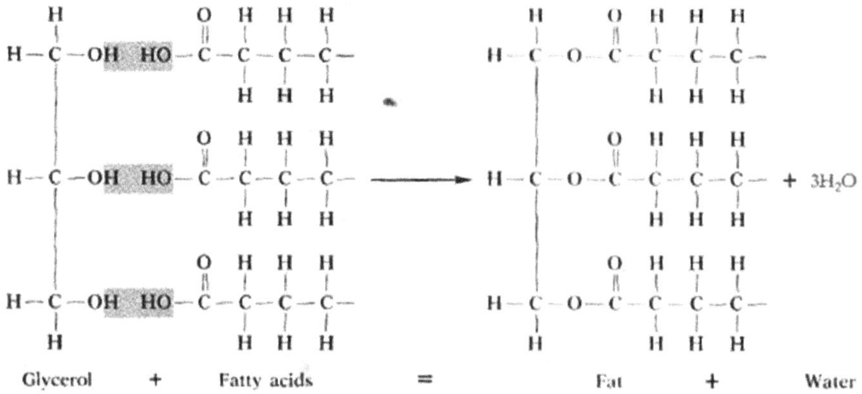

Glycerol   +   Fatty acids   =   Fat   +   Water

Fig. 3.36. The condensation reaction leading to the synthesis of fat is based on the removal of three water molecules and bonding of three molecules of fatty acid to a single molecule of glycerol.

fatty acids and glycerol. Fatty acids have to be absorbed by cells in the digestive tract but lipids are hydrophobic and cells in the digestive track have aqueous environment where fatty acids cannot be solved. In order to overcome the issue of solubility, the fatty acid molecules create, together with phospholipids, spherical structures — the so-called micelles that will be analyzed in Chapter 4 and which are shown schematically in Fig. 3.37.

A membrane prefers non-polar molecules due to the very large hydrophobic region. It also prefers small molecules. Glucose, polar molecules and charged particles cannot pass through the cell membrane without the help of a transporter, which is a protein. The molecule or ion to be transported (substrate) binds at the protein. After crossing the membrane, the carrier protein releases the substrate. In this sense a membrane is selectively permeable.

As mentioned before, diffusion is the process by which molecules move at random and spread from areas of high

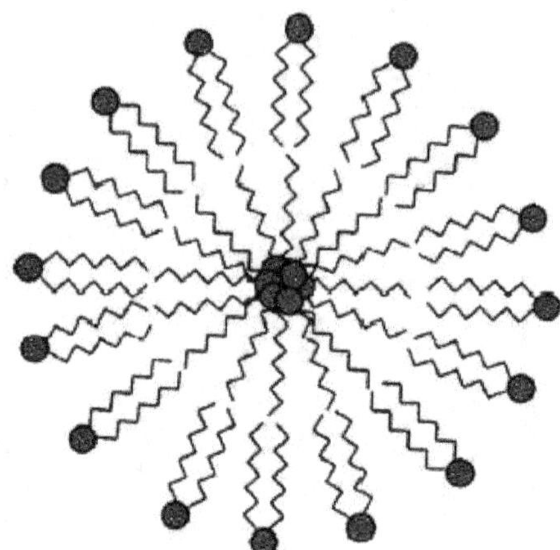

Fig. 3.37. Schematic picture of a micelle composed of an outer shell made of hydrophilic heads and an inner "star" of fatty acids carrying hydrophobic tails. The hydrophilic heads of the outer shell of phospholipids face the aqueous environment.

concentration to areas of low concentration. Water and oxygen move freely across the cell's membrane by diffusion.

3.10.4. *Summary of Properties of Diffusion and Osmosis*

- Spontaneous motion of particles from an area of high concentration to an area of low concentration.
- Net diffusion stops when concentration on both sides becomes equal (if crossing a membrane) or when there is a uniform distribution of particles. When equilibrium is reached, molecules continue to move, but no net change in concentration occurs.
- Smaller particles will diffuse more rapidly than larger particles.
- O_2 and CO_2 rapidly diffuse across the membrane. This is essential for life (breathing process).
- Water, while polar, is small enough to freely move across the plasma membrane.

- Larger hydrophilic uncharged molecules, such as sugars, do not freely diffuse (they can diffuse, but very slowly). They need specific transporters.
- Ion channels and specific transporters are required for charged molecules and larger, uncharged molecules.

CHAPTER 4

Role of Water in Some Fundamental Processes

In this chapter we discuss the role of water in some fundamental biochemical processes, such as protein folding and self-assembly of proteins.

Today we understand that water is not an "innocent bystander". Rather, it is an active agent, and quite a vital one at that. It interacts with biomolecules and takes an active role in processes within the cell to the degree that the water molecule can be looked upon as a biological molecule. That certainly pertains to water molecules near the surface of the protein. There are mutual effects of proteins on nearby water molecules and vice versa. There is a shell of water molecules, the hydrating shell, around the protein. Moreover, that shell is manipulated and shaped by the protein and it determines in multiple subtle ways the functioning of the protein. It determines the shape of the protein including its inner structure, the so-called conformation.

4.1. Distribution of Solutes Between Water and an Organic Liquid

Before we discuss the role of water in some fundamental biochemical processes, let us start with a very simple experiment. Suppose we have a glass with equal amounts of water and say oil, ethanol or benzene, or any other organic liquid. Essentially, in this simple experiment we shall make use of two liquids that do not mix — one is water and the second is some organic solvent, such as hexane, benzene, etc.

Now, add a small quantity of sugar to this two-phase system. Let us assume that we added one gram of sugar. We shall see that most of the sugar will be dissolved in the water, whereas only a little amount will dissolve in the organic liquid. We say that the sugar is *distributed* between the two liquids in favor of the water phase. Figuratively, we say that the sugar molecules *prefer* the water phase because it is *hydrophilic* (hydro — "water", philic — "loving"). Of course, there are no such feelings as love and hatred between molecules, but it is a rather convenient way of expressing the experimental fact — that sugar prefers to be in water — in terms of human attributes. Thus, when we say that a solute, for instance, sugar "loves" water we mean that when distributing itself between water, and say oil, it will prefer to be in the water phase, not all but a majority of it.

Now let us take methane, or ethane or propane. These are small hydrocarbons consisting of carbon and hydrogen atoms. Methane is the simplest one (see Fig. 4.1).

The methane molecule has four carbon–hydrogen single covalent bonds. These covalent bonds are called non-polar covalent bonds because the electrons shared by the adjacent atoms in

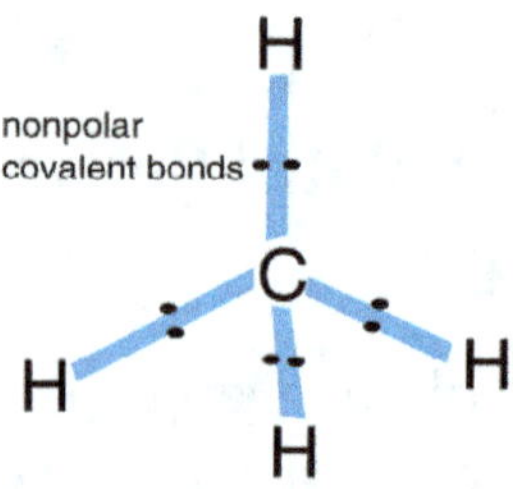

Fig. 4.1.   Methane molecule. Bonds are non-polar covalent.

the bonds are shared equally. The C–H bonds do not have great enough of an electronegativity difference for the bonds to be considered polar. In any case, the T_d symmetry of the methane molecule makes that there is no net (overall) polarity. Methane will therefore be hydrophobic.

We do the same experiment as before, and we will observe that the distribution of methane is in favor of the oil phase. In other words, the majority of the methane molecules will prefer to be in the oil phase, and only a small fraction will be found in the water phase.

Methane is a colorless, odorless gas (it boils at $-161°C$ at a pressure of 1 atm). It is essentially insoluble in water (solubility in water is 23 mg/L at 25°C and 1 atm). As temperature increases, CH_4 solubility decreases.

From an environmental viewpoint, the solubility of methane in water gives rise to different problems. Thus, methane gas can occur naturally in water wells, caused by decomposition of vegetation or other organic materials, mixed in with sediments, thousands or even millions of years ago. Because it is flammable and potentially explosive when mixed with air, it could be dangerous when emitted directly from household faucets. Therefore, methane from the well and water systems should be vented to the atmosphere, outside of enclosed spaces. Removal of methane from water commonly involves aeration (methane is not filterable or chemically removable). On the other hand, a great deal more methane is contained in undersea deposits (clathrates). These are frameworks of hydrogen-bonded water molecules that surround the methane molecules (see Fig. 4.2), and they form solid materials that are stable at sufficiently low temperature and low pressure. Clathrates are found in ocean sediments at depths of 300 meters or more. Extreme care should be exercised when devising methods to extract this methane in order to take advantage of its potential

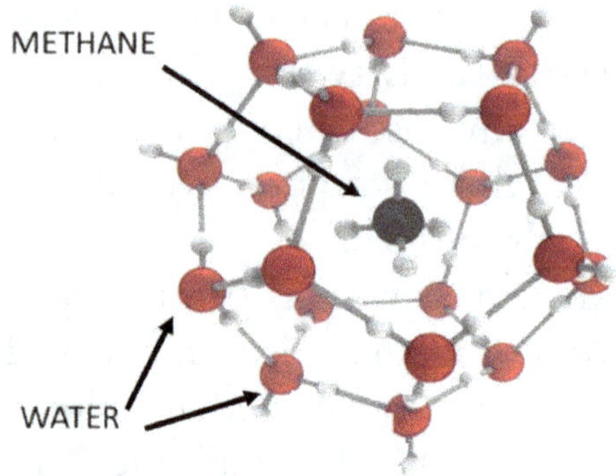

Fig. 4.2.   The structure of a clathrate; a methane molecule surrounded by water molecules.

energy, because its accidental release to the environment would produce important greenhouse effects. Global warming effects on ocean temperature will also contribute in the same direction.

We could say that methane is a "water hater", or an "oil lover". The term that was used for this phenomenon is *hydrophobic* or "*water fearing*" molecule. This is not exactly the antonym of hydrophilic molecule, but the term "hydrophobic" stuck and is widely used throughout biochemical literatures.

You know by now that there are two kinds of molecules called hydrophilic and hydrophobic. Scientists have also suggested a *hydrophobicity scale*. What this means is simple: Take water and oil, and add a third solute. If the ratio of the concentrations of the solute is such that it favors the oil phase, it will be referred to as being more hydrophobic, or less hydrophilic. On the other hand, if we find that the solute favors the aqueous environment we shall say that the solute is hydrophilic.

Note, however, that the scaling of hydrophobicity depends on an experiment, that is, measurements of the relative concentrations of a solute in the two phases. This is purely a macroscopic definition. There is not a hint of why one molecule is more hydrophobic or hydrophilic than the other. The fact is that some solutes are more hydrophobic and some others are more hydrophilic. These facts were well known in the early 1940s. We shall also soon see that the proteins, which are spewed from the ribosomes, must fold before they function. In this process, the protein behaves as if it is a string of "solutes", some hydrophobic and some hydrophilic.

For the purpose of this chapter, we discuss mainly two kinds of solutes: Hydrophobic ($H\phi O$) and Hydrophilic ($H\phi I$). Occasionally we shall also discuss ionic solutes (see Fig. 4.5).

Remember that molecules do not possess such human emotions, but these are only convenient terms that are useful in discussing the role of water in biological processes.

All you have to remember is that a water molecule has four directions along which it can form a hydrogen bond (HB) (see Fig. 4.3). We refer to these directions as "arms" — two H-arms (hydrogen or "hands") and two L-arms (two "legs", each leg

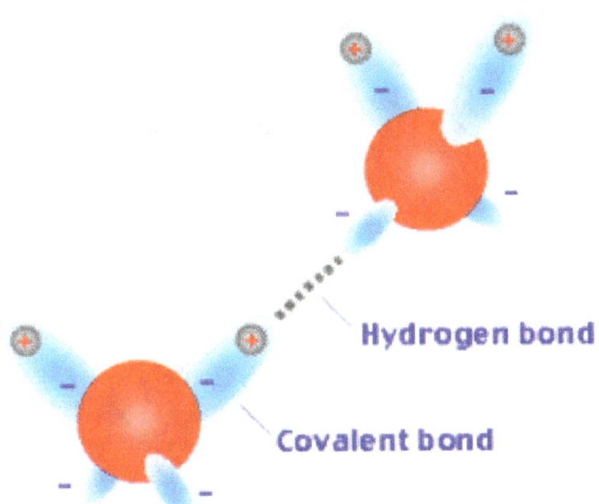

Fig. 4.3. "H-arm or hand" (hydrogen connected covalently to oxygen) and "L-arm or leg" (lone pair of negative charged electrons) in a hydrogen-bonded water–water system.

stands for an electron pair). The H-arm can be attached or be bound to an L-arm. Alternatively, a hand binds to a leg. In reality, the H-arm and the L-arm are termed hydrogen bond *donor* and hydrogen bond *acceptor*, respectively.

A lone pair of electrons exists in other molecules like methanol. You can look at methanol as a water molecule in which one of the hydrogen atoms has been replaced by the group CH_3 (see Fig. 4.4).

Methanol therefore dissolves in water like other alcohols, as will be described later.

For most of the explanation of the properties of water we can disregard the difference between L and H. We can simply view each water molecule as having four arms.

Let us try to understand the molecular reasons for why a solute is hydrophobic or hydrophilic. It depends on the polarity of the solute molecule, or in our language here, on having arms along which they can form hydrogen bonds with water molecules.

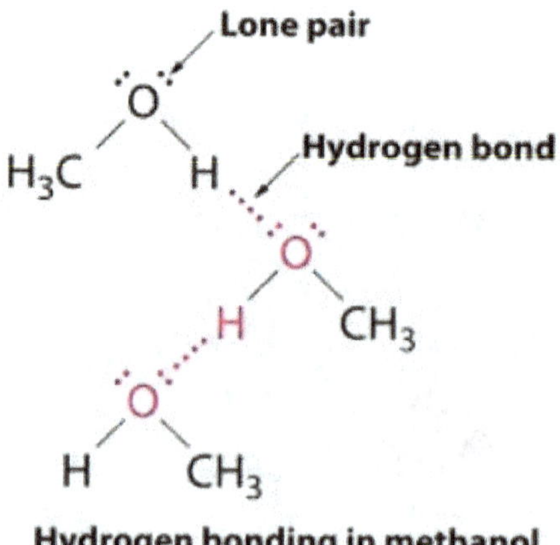

Fig. 4.4.   Methanol–methanol hydrogen bonding interactions. As one can see, a methanol molecule can be viewed as a water molecule in which a hydrogen atom is replaced by a methyl group.

Charged ions resulting from the dissociation of a salt, e.g., NaCl in water, attract the charged end of the water molecule. A Cl⁻ anion attracts the partially positive charged hydrogen atom in water and Na⁺ cation attracts the "leg" of the water molecule, namely the partially negative charged oxygen due to its pair of free electrons as in Fig. 4.5.

Organic molecules may also be polar, e.g., a urea molecule, which has quite a few arms, along which they can form hydrogen bonds with water molecules (see Fig. 4.6).

Hydrophobic molecules have non-polar bonds and are sparingly soluble in water. This property is true, especially in the case of hydrocarbons (see Fig. 4.7), i.e., molecules that are predominately composed of hydrogen and carbon atoms, unlike water that has partial polar oxygen and hydrogen atoms.

The covalent bonds between carbon and hydrogen atoms are non-polar, as in the hydrocarbons methane (Fig. 4.7).

As a reminder: With non-polar covalent bonds, unlike polar covalent bonds, there is an equal sharing of electrons between the carbon and hydrogen atoms. Thus, there is no charge

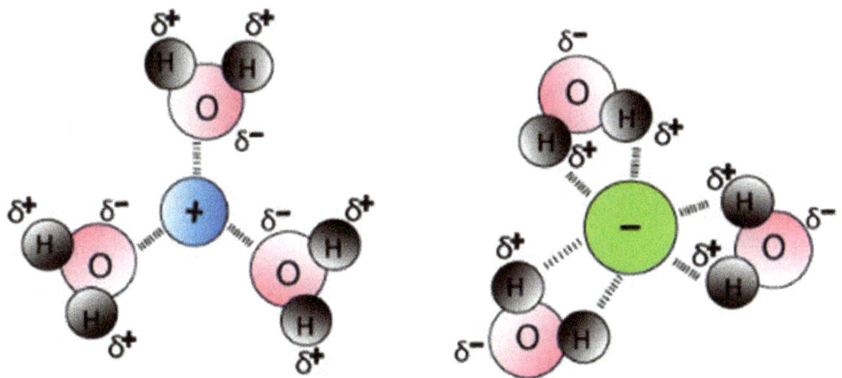

Fig. 4.5. Ionic substances such as salt (sodium chloride) dissolve in water because water molecules are attracted to the positive (Na⁺) or negative (Cl⁻) ionic parts.

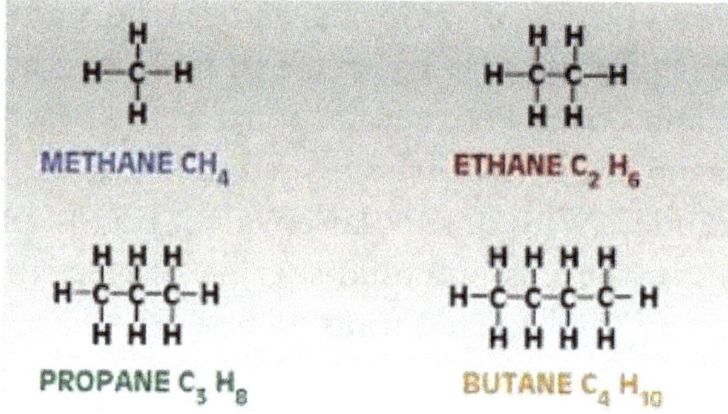

Fig. 4.6.   Polar substances such as urea dissolve because their molecules form hydrogen bonds with the surrounding water molecules.

Fig. 4.7.   The first four members of the hydrocarbon "family".

separation in the molecule and it cannot form hydrogen bonds with water.

Other families of hydrocarbons are the cyclic hydrocarbons and the aromatic hydrocarbons. Figures 4.8 and 4.9 show representative examples of these classes of hydrocarbons: Cyclohexane (C_6H_{12}) and benzene (C_6H_6). Cyclic hydrocarbons are similar to aliphatic hydrocarbons but they form ring structures. In Chapter 2, we mentioned the special bonding in benzene that produces a π electron cloud — six electrons (one on each carbon atom) contribute to the carbon–carbon bonding (see covalent

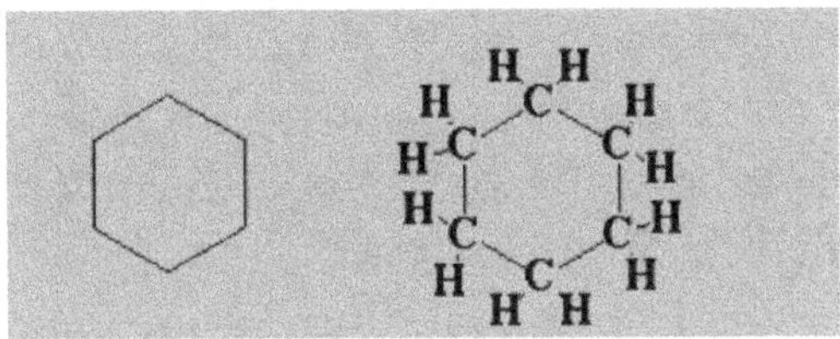

Fig. 4.8.   Chemical structure of cyclohexane (cyclic hydrocarbon).

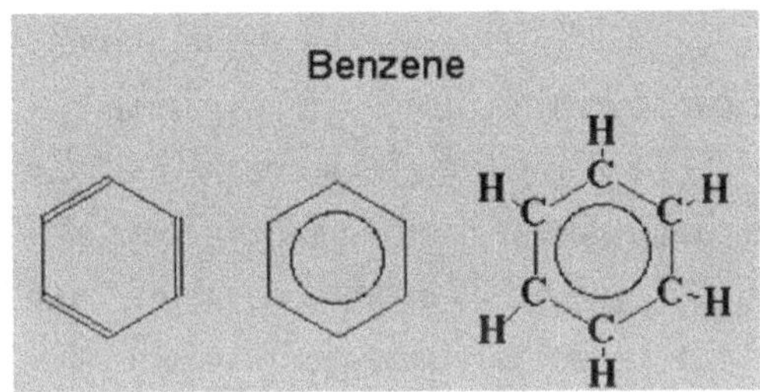

Fig. 4.9.   Chemical structure of benzene (aromatic hydrocarbon).

bonding in Chapter 3), making it stronger than the usual C–C bonding in aliphatic hydrocarbons.

As in the case of the aliphatic members of the family, they are sparingly soluble in water. Aromatic compounds exhibit a slight solubility in water due to van der Waals interactions, as mentioned in Chapter 2.

If a substance X, which is soluble to a greater or lesser extent in two immiscible liquids is added to a mixture and then allowed to settle, the concentrations in each layer become constant. However, there is continual interchange of solute between the liquid layers via the interface, i.e., a dynamic equilibrium is formed.

If more of substance X is added to the system, the solute will distribute itself between the two immiscible liquids so that the ratio of the solute concentrations remains the same at constant temperature independently of the total quantity of X in the same molecular state. This is essentially the *partition equilibrium law*. We can define an equilibrium constant, $K_X = C_X^\alpha / C_X^\beta$, as the quotient between the concentrations of X in each phase (α, β). This is also called the partition coefficient.

Let us take as an example: Iodine (I_2, the molecule of iodine is diatomic) and tetra-chloromethane (CCl_4).

The structure of CCl_4 is similar to that of the methane in which the hydrogen atoms have been replaced by chlorine atoms (see Fig. 4.10). There is a basic difference in regard to polarity. The C–H bonds in methane are non-polar, while the C–Cl bonds are polar. The bonding electron pair is closer to the chlorine atom, which thus exhibits a partial negative charge. The tetrahedral structure of CCl_4 is shown in Fig. 4.10.

It is important to stress that despite the fact that the four bonds are polar the vector sum of the corresponding dipole moments is zero (the resultant four identical vectors with their origin in the carbon atoms and directed toward the vertices of a regular tetrahedron is the null vector).

Both iodine and tetra-chloromethane are non-polar solvents. Therefore, they could mix together. When iodine is dissolved in CCl_4 it changes its color from brown to purple.

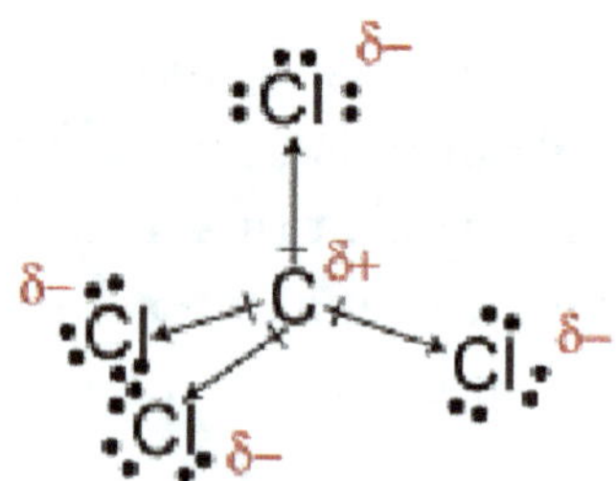

Fig. 4.10. Chemical structure of CCl_4. Note that even though the C–Cl bonds are polar, their symmetrical arrangement makes the molecule non-polar.

The partition coefficient of iodine dissolved in two phases (water and CCl_4) is:

$$K = [I_{2(aq)}]/[I_{2(CCl_4)}] = 0.0116.$$

Iodine is much more soluble in the non-polar organic solvent than in the highly polar water solvent.

The partition coefficient of ammonia dissolved in two phases (water and $CHCl_3$) is:

$$K = [NH_{3(aq)}]/[NH_{3(CHCl_3)}] = 26.$$

The highly polar ammonia can form hydrogen bonds with water molecules, hence ammonia is much more soluble in water than in tri-chloromethane.

4.2. Solubility of Some Molecules

We shall not discuss all kinds of molecules here. Salt solutions were considered in Chapter 3. Here we will focus on molecules that are relevant to the process of protein folding, which will be studied later in this chapter. More specifically, we shall discuss either molecules, which have "arms" to form hydrogen bonds, or molecules that do not form hydrogen bonds.

Now you can see that whenever solutes with one or more arms enter the water, they are "welcomed" by the water molecules. The very fact that these solutes have arms, enables them to "hold hands" with water molecules as if they become "instant friends", thus allowing the molecules to mingle comfortably with the water molecules. On the macroscopic level we observe that these molecules are very soluble in water. Therefore, we have called them "water loving" molecules, or hydrophilic molecules. In laymen's terms, perhaps a better way of describing it is a mutual love affair between these solutes and the water molecules. This is how scientists sometimes talk about molecules, attributing to them feelings of love and hate, which of course they do not possess at all.

4.2.1. *Alcohols*

As will be shown later on in this chapter, the 20 natural occurring amino acids are molecules containing both hydrophilic and hydrophobic groups. Aliphatic groups like H, R-CH$_3$, or aromatic rings like R-C$_6$H$_5$, define the hydrophobic regions because, as discussed in a previous section of this chapter, they are all very little soluble in water. Among the hydrophilic groups we can mention the hydroxyl (R-OH) and carboxyl (R-COOH) groups. The former is present in serine, threonine and tyrosine, the latter in aspartic and glutamic acids.

Let us comment shortly on the solubility of these groups, as it will help rationalize the behavior of proteins in water.

In case of alcohols, just as it happens in the case of many other biological molecules, the basic solubility rule that "like dissolves like" is a bit more complex. Each alcohol consists of a carbon chain (always non-polar) and one or more OH groups (which are polar). Thus, for instance, in the case of butanol, its chemical structure is:

$$CH_3 - CH_2 - CH_2 - CH_2 - OH$$

The structure of an alcohol resembles that of water (see Fig. 4.11).

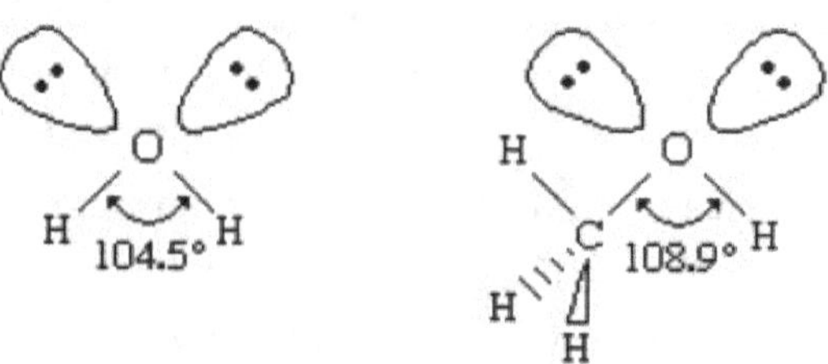

Fig. 4.11.   Chemical structures of water and methanol.

There are alcohols with two OH-groups (ethylene glycol, propylene glycol), three OH-groups (glycerol or glycerin), or more (cholesterol).

Thus, alcohols are compounds that contain the –OH group of atoms attached to a hydrocarbon chain. Methanol, ethanol and propanol are quite soluble in water. Each OH group has three arms (one hydrogen atom and two lone pairs on the oxygen atom). This fact is enough to make these molecules "welcome", hence soluble in water. However, when the chain of the "armless" hydrocarbon becomes larger, the solubility systematically decreases. In fact, even butanol has a limited solubility in water. Pentanol, hexanol, heptanol and others are less and less soluble.

Think of these molecules as having "split personalities" — one side is "water loving" while the other is "water fearing". As the tail of the alcohol becomes larger, the solubility in water will become lower and lower. This fact reminds us of another important phenomenon that is not related to proteins, but which one encounters in one's daily life. Here the interplay between the hydrophobic and hydrophilic groups explains how detergents work (see later on in this chapter). The longer the carbon chain in an alcohol is, the lower the solubility in polar solvents and the higher the solubility in non-polar solvents.

Hydrogen-bonding raises the boiling point of alcohols. This is due to the combined strength of so many hydrogen bonds forming between oxygen atoms of one alcohol molecule and the hydroxyl hydrogen atoms of another. Let us give some details on the simpler alcohols below.

Methanol, also known as methyl alcohol and wood alcohol, is the simplest of the alcohols (see Fig. 4.12). Methanol is found in wood smoke and contributes to the odor of wine. It is metabolized in the body to produce formaldehyde and formic acid and is toxic if more than 50 ml are consumed; smaller amounts can cause blindness.

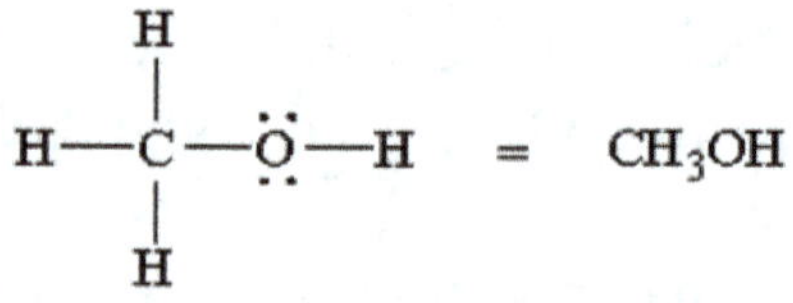

$$H-C-O-H \quad = \quad CH_3OH$$

Fig. 4.12.   Chemical structure of methanol.

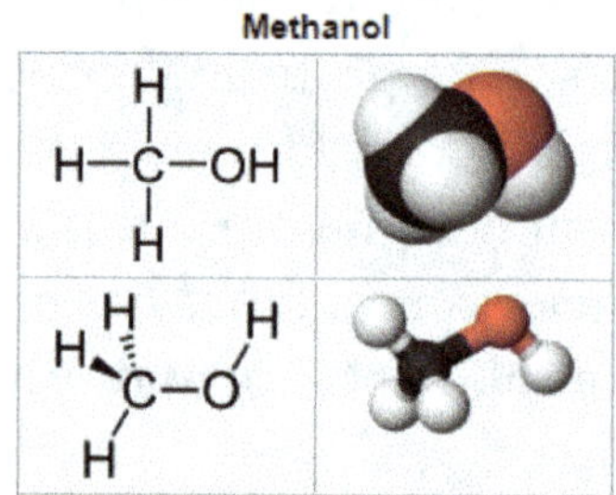

Fig. 4.13.   Methanol.

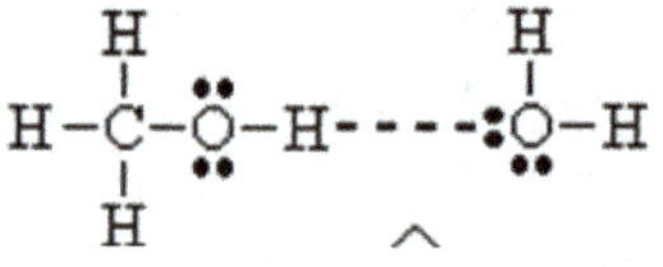

$$H-C-\ddot{O}-H----\ddot{O}-H$$

Fig. 4.14.   Hydrogen-bonding between methanol and water.

Water molecules are attracted to the –OH group of methanol. They make hydrogen bonds between hydrogen atoms in water and the oxygen atom of the –OH group in methanol. They also make hydrogen bonds between the oxygen atom of the water molecule and the hydrogen atom from the O–H group in methanol (see Fig. 4.13).

If you feel more comfortable using chemical formulas, in Fig. 4.14 you can find the hydrogen bond between the –OH group in methanol and the lone pairs in water, Fig. 4.14.

For ethanol, the chemical formula is C_2H_5OH (see Fig. 4.15).

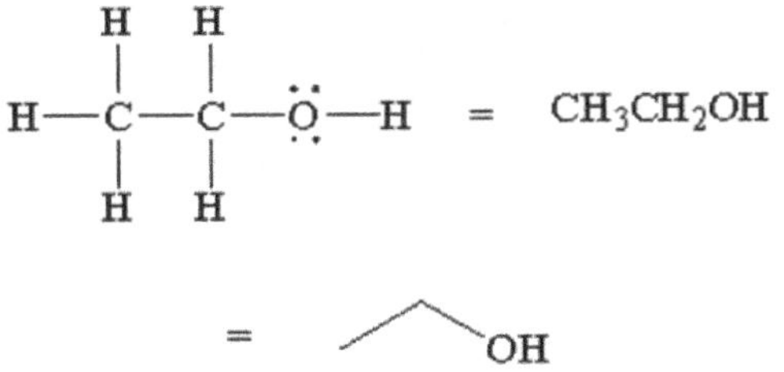

$$H - C - C - \overset{\cdot\cdot}{\underset{\cdot\cdot}{O}} - H \quad = \quad CH_3CH_2OH$$

$$= \quad \diagup\!\diagdown OH$$

Fig. 4.15.   Three different (equivalent) chemical representations of ethanol.

Ethanol has a two-carbon chain and a –OH group. As water is polar it attracts the –OH group. The carbon chain on the other hand is non-polar and it is repelled by water. The solubility of ethanol is therefore determined by the stronger of these two forces.

Ethanol, also known as ethyl alcohol and *grain alcohol*, is the alcohol found in alcoholic beverages. The fermentation of the sugars found in honey, grain and fruit juices by yeasts to yield beers and wines was probably the first chemical reaction to be discovered.

Ethanol can be purified and concentrated by fractional distillation, but ethanol and water form a constant-boiling mixture at a concentration of 95% ethanol and 5% water, which cannot be separated by distillation. Chemists call this mixture an azeotrope. An azeotrope cannot be further separated by distillation because when boiled the vapor has the same proportions of ethanol and water as the un-boiled mixture. Absolute ethanol, which contains no water, is produced by fractional distillation of 95% ethanol with small amounts of benzene. Ethanol that is intended for industrial use is "denatured" (rendered unfit for human consumption) by adding small amounts of methanol or other unpleasant or toxic substances.

Ethanol is metabolized in the body, primarily by the enzyme alcohol dehydrogenase, to produce acetaldehyde; the buildup of acetaldehyde in the blood is one of the factors that contribute to the symptoms of a hangover. Physiologically, ethanol acts as a depressant, but since it frees parts of the cortex from inhibitory controls, it seems to be a stimulant to its consumer.

The third alcohol in the series, which has a three-carbon alcohol with the –OH group on an end carbon, is called 1-Propanol, or propyl alcohol.

1-Propanol is formed naturally in small amounts during many fermentation processes and is used as a solvent in the pharmaceutical industry.

Because of the strength of the attraction of the –OH group, the first three alcohols — methanol, ethanol and propanol — are completely miscible with water; they dissolve in water in any amount. From the four-carbon butanol onwards, the solubility of alcohols starts to decrease. After the 7-carbon heptanol, alcohols are considered insoluble in water. As the number of carbon atoms in the chain increases the solubility of alcohols in water decreases. We can see this in the measured solubilities listed below:

Alcohol Solubility in Water [g/100 mL water]

Methanol	Unlimitted
Ethanol	Unlimitted
Propanol	Unlimitted
Butanol	8.0
Pentanol	2.2
Hexanol	0.7

4.2.1.1. *Antifreeze*

An antifreeze is a substance added to water to avoid it freezing when temperatures are below 0°C (freezing would cause pipes to clog and burst because of the expansion of water upon freezing).

We add salt to icy roads, and we put antifreeze (a mixture of ethylene glycol and water) in our car's radiators to prevent them from freezing. Also the cells of some organisms produce antifreeze proteins to prevent lethal formation of ice crystals. All these processes take place because a freezing-point depression occurs.

The extent of the freezing-point depression is based on two factors: (a) The number of dissolved particles present, and (b) the nature of the dissolved particle. Sometimes the solutions are also named antifreeze. As we already mentioned, alcohol molecules interact with water molecules through –OH groups, forming hydrogen-bonded structures. Thus, they inhibit at some degree the water–water hydrogen-bonding interactions leading to the formation of ice. This resulting inhibition allows water and water solutions to get supercooled, without ice formation, to temperatures below the thermodynamic freezing point.

Glycols are the main commercial antifreeze solutions. Figure 4.16 shows the chemical structure of two glycols. It is important to stress that the antifreeze solution also raises the boiling point. For a 50/50 mixture, the boiling point increases from 100°C (water) to about 107°C.

Why does an ethylene glycol-water mixture have a higher boiling point? The boiling point of a solution is the temperature at which the vapor pressure of the solution equals the atmospheric pressure. As more of the less volatile substance (ethylene glycol) is present, there would be fewer molecules escaping into the vapor phase. In order to keep the same vapor

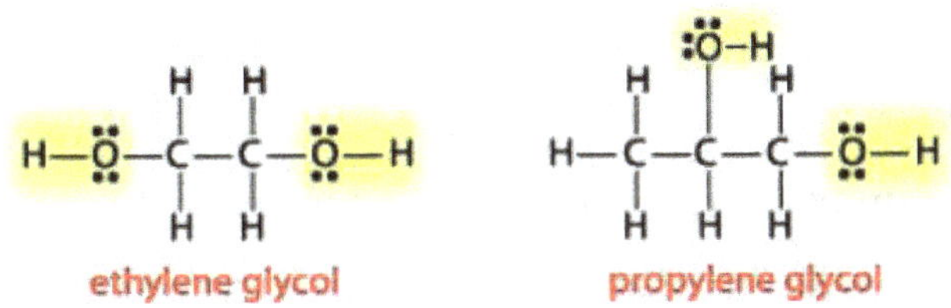

Fig. 4.16.   Chemical structures of two glycols.

pressure, more heat (higher temperatures) would be required. This could be a problem as most engines operate at around 90–105°C and it is important to keep the antifreeze solution from boiling, because boiling antifreeze does not do its job very well. Fortunately, we can control this situation by just increasing the pressure, as we know that increasing pressure increases the boiling point of a liquid. In this context, we should not worry about the composition of our antifreeze solution in a pressurized system.

Another critical aspect from a cooling viewpoint refers to the specific heat capacity of the antifreeze solution. That is to say, how much heat energy the coolant can carry out from the engine to the radiator. Pure water is much more efficient than the antifreeze solution because of its higher heat capacity. However, water has a notable lower cooling capacity than the antifreeze solution. So the best decision is to use an antifreeze solution with an appropriate composition to optimize cooling and heat capacity capabilities.

Commercial antifreeze (the undiluted liquid) is usually 95% weight ethylene glycol with some anti-foaming, anti-oxidant and colorants. Antifreeze mixture (i.e., antifreeze solution) is prepared with nearly 50/50 water-antifreeze. One such a mixture has a freezing point of −37°C (see Fig. 4.17), where the melting points of ethylene glycol-water mixtures are represented. There are other natural antifreeze substances; besides alcohols, sugars and salts, there are some antifreeze proteins that may keep cells alive down to −15°C.

4.2.2. *Sugars*

Let us summarize the difference between what we called hydrophobic and hydrophilic solutes. The first are those that are "rejected" by the water molecules, while the second are very much "accepted". Experience shows that common sugar

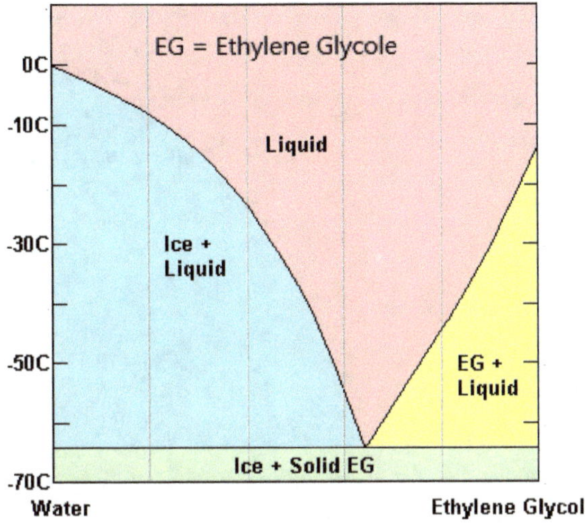

Fig. 4.17.   Ethylene glycol-water phase diagram.

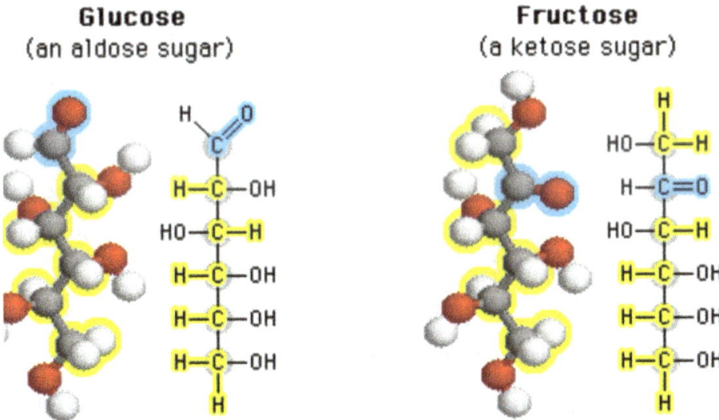

Fig. 4.18.   Glucose and fructose are monosaccharides; these are simple sugars. Aldose means that the sugar contains one aldehyde group (–CH=O).

is a very soluble molecule. The reason is simply that a sugar molecule has many hands (see Fig. 4.18), so it can attach to many water molecules simultaneously. Let us have a look.

Figure 4.18 contains two representative examples of monosaccharides, a kind of sugar.

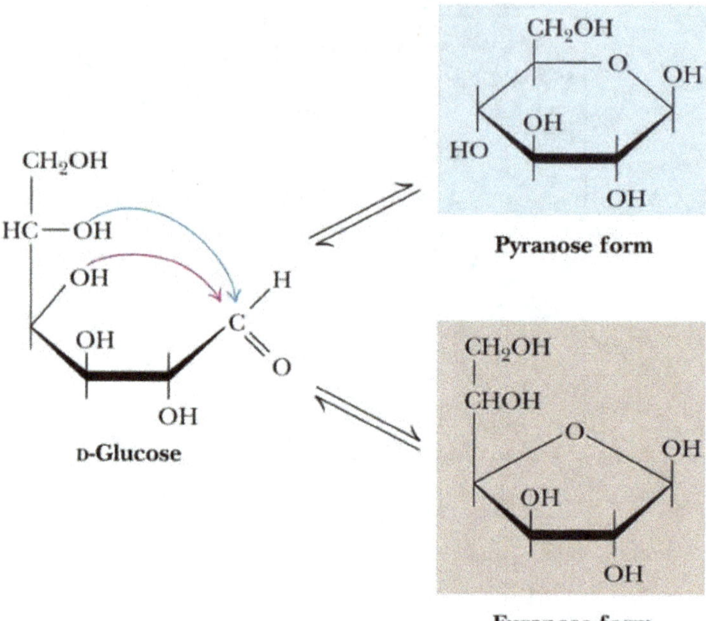

Fig. 4.19.   D-Glucose (open structure) in equilibrium with two closed structures: Pyranose and furanose.

You will hear a lot about those two kinds of sugars concerning metabolism diets and food in general. Monosaccharides usually switch from the open-chain form (see Fig. 4.18) to a cyclic structure (see Fig. 4.19), thus giving rise to an equilibrium between the two forms. For many monosaccharides, the cyclic forms predominate. Thus, glucopyranose forms of glucose predominate in solution.

We will also mention sucrose, which is the common sugar we buy in food shops. Unlike glucose and fructose, sucrose is a disaccharide composed by glucose and fructose (see Fig. 4.20).

Sucrose dissolves more easily in a freshly poured cup of tea than in a cup of tea that has been cooled after left standing for a while. The solubility is, in general, temperature dependent. The following chart (see Fig. 4.21) shows the solubility of

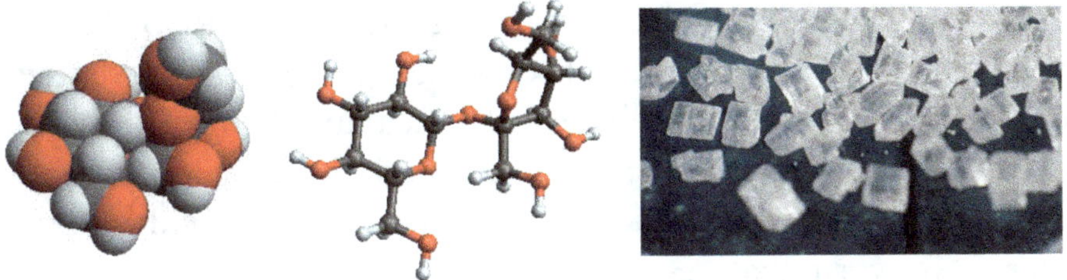

Fig. 4.20.   Chemical structure of common sugar (sucrose).

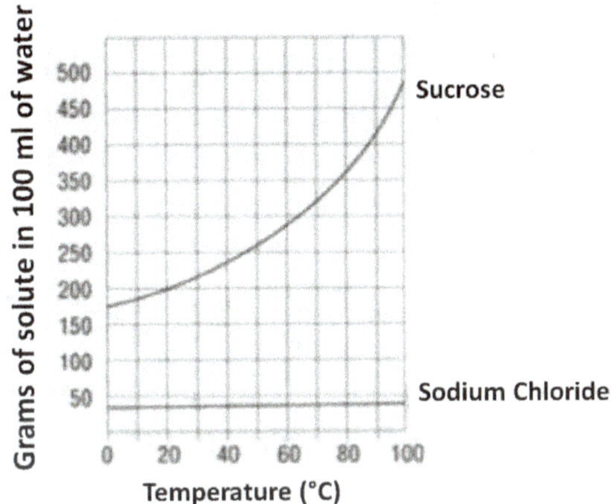

Fig. 4.21.   Solubility of common sugar and a common salt in water.

sugar compared with that of a salt in water as a function of the temperature.

A general rule establishes that most solids increase their solubilities with temperature. In most cases, we can associate this behavior with the fact that the dissolving process is endothermic. Consequently, Le Chatelier's principle tells us that a temperature increase would lead to an equilibrium shift towards more solubility. However, solubility is a very complex subject, which involves many factors. In the case of a solution of gases in water, the effect of temperature is such that as temperature increases the solubility decreases. The only thing you have to do to observe this behavior is to heat a can of soda. You will immediately notice a flow of small bubble of gases leaving the liquid phase.

The impact of increasing the temperature of water on fish in the sea can be dramatic, as the required oxygen will become scarcer in warm waters.

Pressure has little effect on the solubility of solids and liquids in water but it plays a major role in the case of gases. The so-colled Henry's law, in honor of the English chemist William Henry, who studied the subject in the early 19th century and found that solubility is directly proportional to the partial pressure of the gas in equilibrium with the solution. Thus, for instance, the solubility of carbon dioxide in water changes from $1.3 \cdot 10^{-5}$ mol/L at 25°C and 1 atm (with a partial pressure of CO_2 in the atmosphere of about $3.9 \cdot 10^{-4}$ atm) to 0.1 mol/L at 25°C and 4 atm (approximatlly the pressure of CO_2 inside of a bottle of sparkling mineral water).

Pure water can only hold up to 60% weight of sugar in solution at 0°C, growing to 80% weight at 100°C (e.g., at 15°C, up to 1.6 kg of sugar can be dissolved in 1 kg of pure water, forming a saturated solution containing 790 kg/m^3 of sucrose and 500 kg/m^3 of water). Metastable and supersaturated solutions might allow up to 2.3 kg of sugar to be dissolved with 1 kg of pure water at 15°C. Most used syrups have some 40% weight of sugar. It is important to apply an effective stirring for dissolving (better done on heating); otherwise, water over a layer of sugar would take days to reach the equilibrium concentration.

4.2.3. *Carboxyl acids*

A carboxylic acid is an organic compound that contains the carboxylic group –COOH (see Fig. 4.22). The carbon atom is bonded to an oxygen atom by a double bond (stronger than a

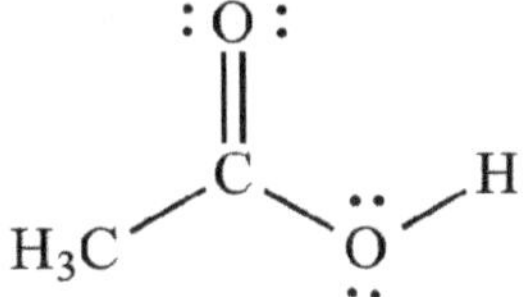

Fig. 4.22.   Chemical structure of acetic acid.

single bond; the C=O distance is appreciably shorter than the C–O distance) and to a hydroxyl group (–OH) by a single bond. A fourth bond links the carbon atom to a hydrogen atom or to some other univalent combining group. Each oxygen atom holds two pairs of non-bonding electrons.

Carboxylic acids are polar substances. Both oxygen atoms attract electron charge from the C=O and C–O bonds, as an oxygen atom is more electro-negative than both carbon and hydrogen atoms. Therefore, relatively strong permanent dipoles are created. The O–H group is even more strongly polarized than the O–H group of alcohols due to the presence of the adjacent carbonyl moiety. The partial charges present in carboxylic acids allow these compounds to participate in energetically favorable hydrogen-bonding interactions with water. As a direct consequence, carboxylic acids with small hydrocarbon chain length ($HCOOH$, CH_3COOH, CH_3CH_2COOH and $CH_3CH_2CH_2COOH$) are soluble in water. As the chain length increases, solubility decreases from $CH_3\text{-}(CH_2)_{10}\text{-}COOH$ onwards, they become insoluble in water as in the case of alcohols (see Table 4.1).

It is important to note that the solubility of carboxylic acids in water does not only depend on dipolar interactions. Other factors, like the ionization in aqueous solutions of appropriate pH, may contribute.

The dipole structure of carboxylic acids favors the formation of a unique hydrogen-bonding arrangement (see Fig. 4.23). Two

Table 4.1.   **Properties of Some Carboxylic Acids.**

Acid's Name	Formula	Melting (°C)	Boiling (°C)	Solubility
Methanoic	CH_3COOH	8	101	miscible
Ethanoic	CH_3COOH	17	118	miscible
Butanoic	$CH_3(CH_2)_2COOH$	−6	164	miscible
Hexanoic	$CH_3(CH_2)_4COOH$	−3	205	10
Decanoic	$CH_3(CH_2)_8COOH$	31	269	2
Hexadecanoic	$CH_3(CH_2)_{14}COOH$	63	269	insoluble
Octadecanoic	$CH_3(CH_2)_{16}COOH$	70	287	insoluble

Fig. 4.23.   Dimer interaction through a double hydrogen bond of two carboxylic acids.

carboxylic acid molecules are held together by two hydrogen bonds. It is a rather stabilizing interaction that is present even in the gas phase. In the liquid phase, a mixture of hydrogen-bonded dimmers and higher aggregates are present, being responsible for the relatively high boiling points exhibited by these compounds.

As in the case of alcohols, organic compounds having multiple carboxylic acid functional groups exist. Thus, for example, oxalic (COOH-COOH) or maleic (COOH-CH_2-COOH) acids have two carboxylic groups.

Carboxylic acids are very common in Nature. For examples, ants use formic acid as a weapon, acetic acid is present in wine when a bottle is left open for a while, lactic acid is formed in muscles during exercise, and malic and citric acids contribute to the tart taste of many fruits and vegetables.

4.2.4. *Amino acids*

Amino acids are carboxylic acids that contain an amine group and a unique organic side chain. From among the natural occurring amino acids, there are 20 that are required for human life to exist. Of those 20 amino acids listed in Fig. 4.24, nine (Histidine,

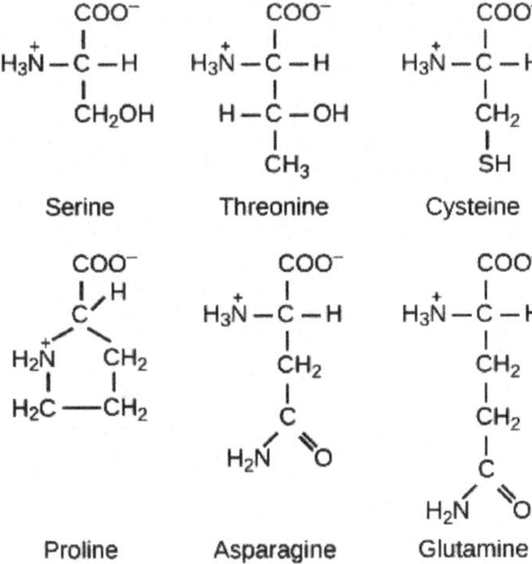

Fig. 4.24.   The 20 natural occurring amino acids.

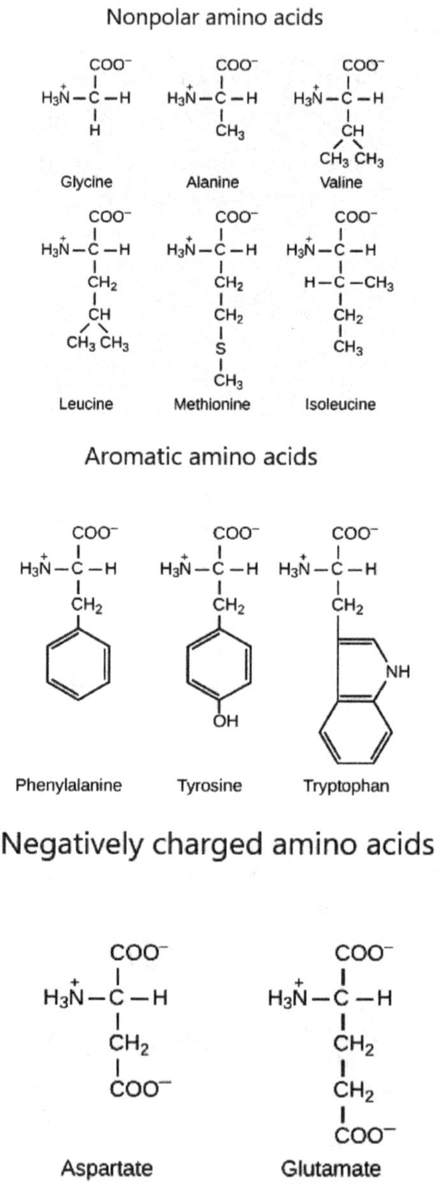
Nonpolar amino acids
Glycine
Alanine
Valine
Leucine
Methionine
Isoleucine
Aromatic amino acids
Phenylalanine
Tyrosine
Tryptophan
Negatively charged amino acids
Aspartate
Glutamate

Fig. 4.24. (*Continued*)

Isoleucine, Leucine, Lysine, Methionine, Phenylalanine, Threonine, Tryptophan and Valine) are essential because they cannot be synthesized from scratch by the human organism, and they must be supplied through one's diet.

Thousands of combinations of those 20 amino acids are employed to build up all the proteins of the human body. They are usually grouped according to the characteristics of their side chains. Thus we have the aliphatic (for example, alanine or valine), aromatic (for example, tryptophan), acidic (for example, aspartic acid), basic (for example, histidine), hydroxylic (for example, serine), and sufur-containing (for example, methionine) or amidic (for example, glutamine).

Amino acids are crystalline solids with high melting points. They tend to decompose before they melt. Decomposition and melting are usually in the 200–300°C range. The notable high melting points of amino acids are associated to their zwitterionic form, even in the solid state. A zwitterion is a neutral molecule exhibiting two local apposite charged groups, namely, $-NH_3^+$ and $-COO^-$. They give rise to strong ionic interactions between one molecule and its neighbors.

Amino acids are generally soluble in water (hydrophilic solutes) and insoluble in non-polar organic solvents such as hydrocarbons. Again the zwitterionic interactions are responsible for the solubility in water. It also depends on the size and nature of the side chains. The presence of hydrophilic groups like $-OH$, $-COOH$, $-CONH_2$, $-NH$ or $-NH_2$ in the side chains contributes to the solubility, as it has been discussed in the preceding sections.

Another important property of amino acids is their optical activity (see Fig. 4.25). The lack of a plane of symmetry (the two mirror images of an amino acid are non-superimposable) produces the existence of two stereoisomers (glycine is the exception). The α-carbon (grey atom in Fig. 4.25) is said to be a

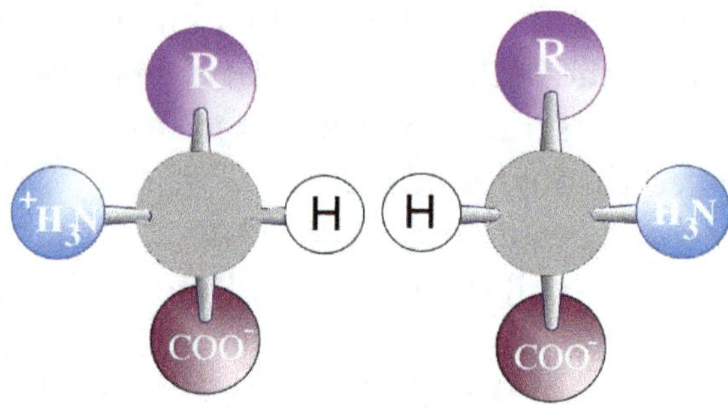

Fig. 4.25.   Optical isomers (enantiomers) of an amino acid. The two enantiomers are mirror images.

chirality center. When the amine group is at the left keeping the carboxyl group at the top, as in the left image of Fig. 4.25, the amino acid is said to have the L configuration at its α-carbon atom. The alternative configuration (see the right image in Fig. 4.25) is the D configuration.

If an enantiomer rotates the light clockwise (as seen by the viewer towards where the light is traveling), that enantiomer is named dextrorotatory and is labeled (+). Its mirror image (levorotatory) is labeled (−). The (+) and (−) isomers are unrelated to the D/L labeling. The latter does not indicate which enantiomer is dextrorotatory or levorotatory.

An important point from a biological viewpoint is that DNA only codes for L-amino acids.

Let us now examine the acid-base behavior of amino acids in certain detail. Let us focus on the particular case of glycine. Figure 4.26 shows the titration curve of glycine.

Glycine contains an acid functional group ($-NH_3^+$) and a basic functional group ($-COO^-$). It is an amphoteric molecule. In a strongly acidic medium (pH $= 0$), the species present is $^+H_3N\text{-}CH_2COOH$ (see Fig. 4.26).

As the pH is raised (addition of a base to the solution), a proton is recovered from this species. The ammonium ion (NH_4^+) is

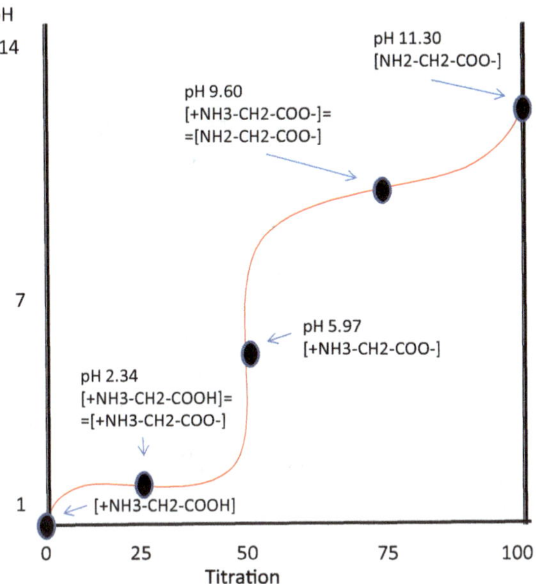

Fig. 4.26.   Titration curve of glycine.

less acidic than a carboxylic acid (CH_3COOH). Therefore, the first deprotonation takes place in the carboxylic group:

$$^+H_3N\text{-}CH_2COOH(\text{-}H^+) \rightarrow {}^+H_3N\text{-}CH_2COO^-, \quad pK = 2.34.$$

When the number of moles added of OH^- equals the number of moles of H^+ originally present, we reach the equivalent point (pH $= 5.97$) where the concentration of the zwitterions is at a maximum. It is called the isoelectronic point or isoionic point, because the molecule has no net charge. Further addition of moles of OH^- leads to the second deprotonation:

$$^+H_3N\text{-}CH_2COO^-(\text{-}H^+) \rightarrow H_2N\text{-}CH_2COO^-, \quad pK = 9.60,$$

producing a second buffer solution at pH $= 9.60$.

Two amino acids react to form a dipeptide with a peptide bond (see Fig. 4.27). New amino acids can be added, thus

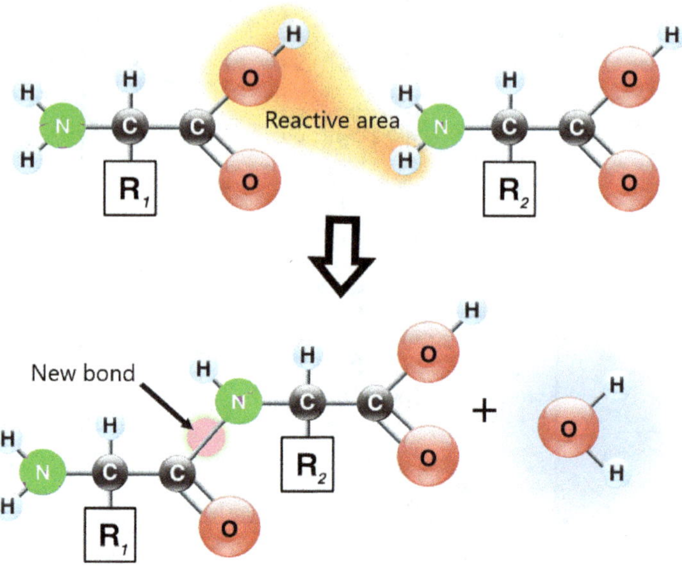

Fig. 4.27.   Dipeptide formation.

forming a polypeptide. Proteins are polypeptides that have some biological functions.

4.3. Micelles and Detergents

Why does soap clean better than just plain water?

In the series of alcohols that we saw before, when the hydrocarbon chain becomes long enough the molecules are welcomed by water from one side and repelled from the other side. This phenomenon is at the heart of the activity of various detergents or soaps. All of these molecules have double faceted personalities — one "loves" water, while the other "fears" water and loves fat. These molecules are called amphiphilic molecules (in Greek this means: Amphis — "both", philia — "love"). They exhibit both hydrophilic ("water loving") and lipophilic ("fat loving") properties (see Fig. 4.28).

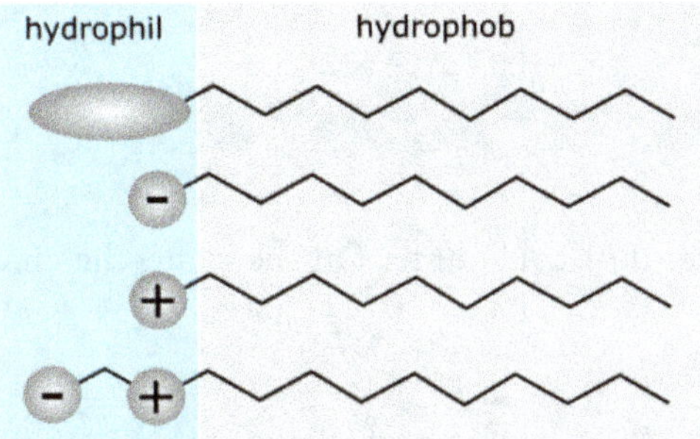

Fig. 4.28.   Four types of surfactant molecules are either ionic or polar covalent as an –OH group in alcohols.

Amphiphilic molecules are formed when we have a long chain molecule, which in itself is very insoluble in water but contains at its edge a hydrophilic group such as an alcohol group, –OH, or a carboxylic acid group –COOH. This happens because these molecules are sometimes partially welcomed in the water, and the hydrocarbon part tends to be expelled out of the water, facing the air or the oil phase.

Organic compounds, which are amphiphilic, meaning that they contain both hydrophobic (their tails) and hydrophilic (their heads) groups, and are called *surfactants*. Thus, a surfactant contains both a water insoluble (or oil soluble) component and a water soluble component. The origin of its name comes from the combination of the surface and agent.

A micelle is an aggregate of surfactant molecules dispersed in a liquid.

An important molecule of that kind that welcomes water only at its edge is called detergent (soap). Figure 4.29 shows its structure.

Its name is sodium dodecyl sulfate (SDS) and it has a negative charge associated to the sulfate group. Its detailed formula is:

$$CH_3(CH_2)_{10}CH_2O-\overset{\displaystyle O}{\underset{\displaystyle O}{\overset{\|}{\underset{\|}{S}}}}-ONa$$

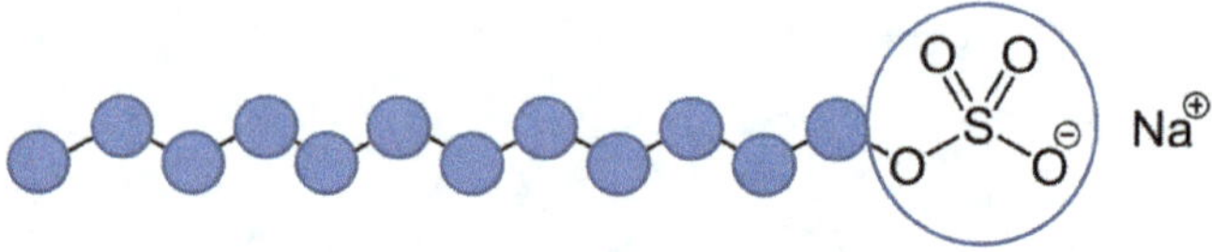

Fig. 4.29.   Sodium dodecyl sulfate. On the right, the charged end enables solvation.

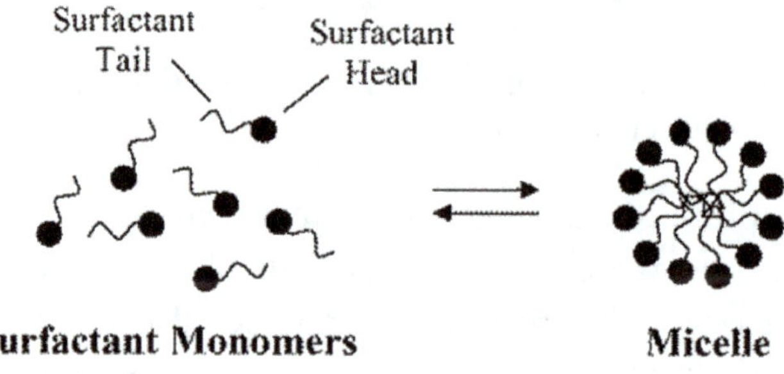

Fig. 4.30.   Schematic illustration of micelle creation.

Micelles are important to the living cell, to chemical engineering, and to pharmacology.

Can a molecule be expelled from water and yet stay in water at the same time? This seems to be a paradox.

This phenomenon is called *micellization* (see Fig. 4.30). As you can see here, these molecules aggregate in such a way that all the "water hating" parts gather at the center of a sphere, while the surface of the sphere is made up of the hydrophilic groups. Large molecules, having a hydrophobic tail and a hydrophilic head, gather together to form a huge sphere; its insides contains all the "tails" while the surface of the sphere consists of all the "heads" (see Fig. 4.30).

The whole micelle is in the water. Why? Because the micelle, as a whole, exposes to water only those groups that can hold hands with water molecules. At the same time all the hydrophobic groups are in the very interior of the micelle and therefore have no contact with water. So, from the point of view of the water molecules, they welcome the micelles into their neighborhood, "unaware" that what lies hidden underneath the seemingly "water loving" exterior are, in fact, all the hydrophobic groups. These groups are not expelled from the water phase, but they adopt a conformation that prevents their "water hating" groups from being in contact with the water molecules. Now, you can understand what we meant by "molecules being expelled from water but remaining in water".

4.3.1. *The Action of Detergents*

This is really a fascinating phenomenon — the long chains are expelled from water, but yet still remain in the water. In order to *please* the water molecules and not to *offend* them, these molecules aggregate in such a way that only their hydrophilic "heads" remain afloat in the water, keeping their hydrophobic "tails" segregated away from the water. So in a certain way these molecules are both "in" and "out" of the water.

We just explained that detergent molecules are like long eels with a few hands on their heads that are welcomed by water. On the other hand, their tails are not welcomed so they aggregate inside the micelles. Now, you can see the whole micelle as a bubble made up of two very different parts — a hydrophilic surface that happily mingles with water and an interior that is more like a drop of oil. You will probably ask what this has to do with the cleaning action of soap? Here is the answer below.

When you have stains (either on your clothes or on your hands) that are not soluble in water, say a drop of oil, the micelle can absorb this oil molecule into their interior. If you wash your hands using plain water, these oil drops will not dissolve. Remember, oil is "hydrophobic" and water would not like to carry it. On the other hand, when you wash your hands or clothes with detergent, some oil-like stains will be *absorbed* by the micelles interior. This action is like dissolving oil with other oil. However, in water, micelles as a whole can carry away the oil stains with it and therefore clean your hands. This phenomenon is called *solubilization*, meaning that micelles, in their peculiar structure can dissolve hydrophobic molecules. Since the whole micelles are in water, the net effect is that water is now enabled to also dissolve hydrophobic molecules. These molecules by themselves are not soluble in water, but they are "*solubilized*" in the presence of micelles, see Fig. 4.31.

The building stones of the cell membrane are an example of split personality molecules — hydrophobic at one end and hydrophilic at the other. This enables transfer in and out of the cell membrane, as needed by the cell.

Proteins and lipids are the main components of cell membranes. In the case of phospholipids (membrane lipids), the end

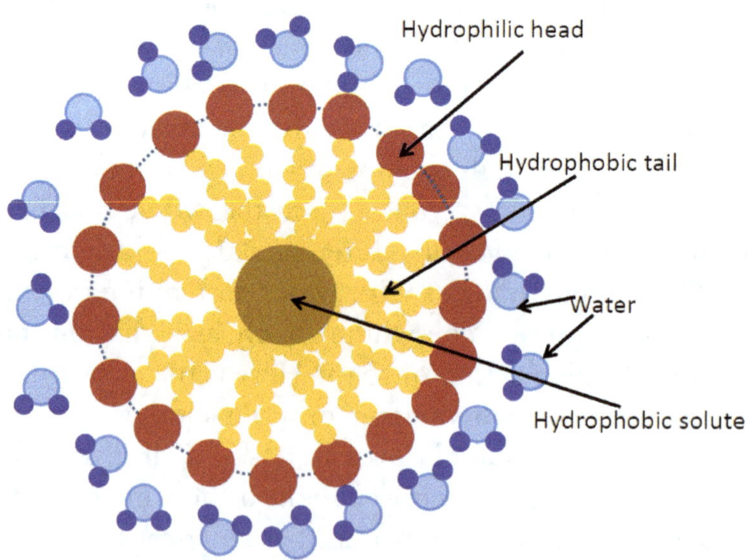

Fig. 4.31.   Solubilization of a hydrophobic solute in the micelle.

of the molecule contains a "water loving" group, while the other end is a "water fearing" long hydrocarbon chain. Lipids form a bilayer with the polar groups facing the extracellular space as well as the lumen of cell and the non-polar groups buried in the interior region. Proteins can be found close to the membrane surface or they can even penetrate partially or totally through the membrane. The extent of interaction between the proteins and lipids would depend on the types of intermolecular forces and thermodynamic considerations. The role played by these proteins is threefold: They (a) constitute the structural support of the membranes, (b) speed up the catalytic effect acting as enzymes, and (c) actively transport ions and other molecules across the membrane.

Micelles have *particular significance* in the pharmaceutical industry because of their ability to increase the solubility of sparingly soluble substances in water.

Micelles have an anisotropic water distribution within their structure. The water concentration decreases from the surface towards the core of the micelle, with a completely hydrophobic (water-excluded) core. The spatial position of a solubilized drug in a micelle will therefore depend on its polarity: Non-polar molecules will be solubilized in the micellar core, and substances with intermediate polarity will be distributed along the *surfactant* molecules in certain intermediate positions.

Micellar solubilization is an efficient alternative for dissolving hydrophobic drugs in aqueous environments. The use of micelles in pharmacy finds numerous applications. Micelles have the preferred role of a drug carrier compared to other alternatives as a soluble polymer. The application of micelles in drug delivery, in order to minimize drug degradation and loss, enables drugs to stay in the blood for longer periods, providing gradual accumulation in the required areas and preventing harmful side effects.

Finally, we mention here the phenomenon of inverse micelles.

In a micelle, water can be seen as surrounding an organic micro-droplet. In an inversed micelle, liquid water exists as a micro-droplet, a kind of a mini pool inside organic solvents (e.g., benzene).

The droplet (typically 4–25 nanometers) is surrounded by surfactants, in between the water droplet and solvent. The polar or ionic charge head group are directed to the water pool while non-polar tails are oriented to the outside, towards the organic solvent.

4.4. The Role of Water in Protein Folding

As we emphasized previously, studying water and its structure and dynamics is a most exciting research topic in modern science. The same can be said about the functionality of the protein, which is no less an intriguing topic of research.

Proteins, the workhorses of living systems, are constructed from chains of amino acids, which are synthesized in the cell based on the instructions written in the DNA. Once a protein is formed, it must fold into a specific structure in order to function as a "working protein". The time for folding varies from microseconds to hours. In this section we shall first discuss the protein-folding problem, then we shall focus on the critical role of water in shaping and structuring the protein and in determining the speed of the folding process.

4.4.1. *The Structure of Proteins*

A protein is a sequence of amino acids. This sequence defines the so-called primary structure of the protein.

To become functional, the protein chain has to get a particular 3D form (tertiary structure), one among the enormous number of possible shapes that is needed for its specific function in the living cell.

This occurs through an intermediate form, known as the secondary structure, the most common of which are the helical structure (the so-called α-helix) and the plate-like β-pleated sheet chain that folds over on itself (see Fig. 4.32).

If several amino acid chains assemble to form a given protein, its global view gives you what is called the quaternary structure.

Although it is possible to deduce the primary structure of a protein from a gene sequence, its tertiary structure cannot be determined. It is therefore not clear how an amino-acid chain folds into its tertiary structure in the short time scale (fractions of a second).

Proteins are constantly restless. They are always on the move — fluctuating, folding and unfolding. They keep doing this even at very low temperature and this constant changing and motion is intimately coupled with the motion of water molecules around them.

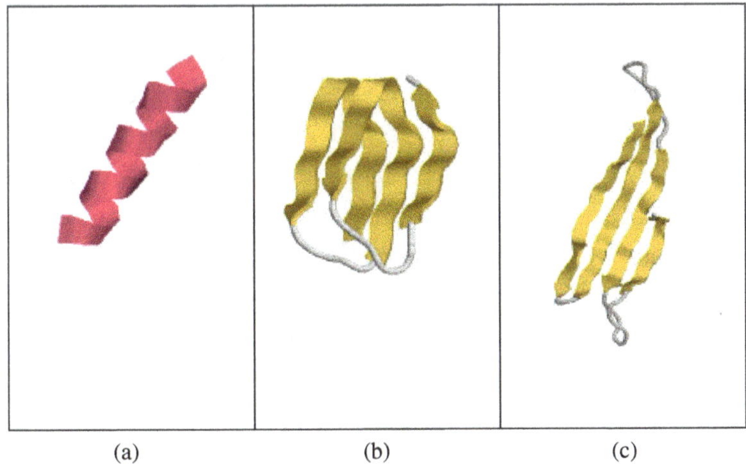

(a)   (b)   (c)

Fig. 4.32.   Examples of protein secondary structures: (a) α-helix, (b) parallel β-sheet, and (c) anti-parallel β-sheet.

A protein has two basic structures critically determining its biological functioning: The folded "native" state and the unfolded ("denatured") state. The transition of the protein between those two states is reversible.

As described previously, proteins are very long strands of amino acids linked together in specific sequences. The protein strand, once created, begins to fold spontaneously. Sometimes enzyme assistance is needed. These are referred to as chaperones.

It is known that while in the folded state most of the hydrophobic elements of the protein are buried deep inside the structure and hydrophilic elements end up on the outside, i.e., are exposed to the water. Clearly, the water environment is crucial to maintaining the stability of the structure of the protein.

4.4.2. *The Protein Folding Problem*

One of the most challenging problems in molecular biology is the so-called Protein Folding Problem (PFP). In order to function, the protein must first fold into a very precise 3D structure (not all proteins have a well-defined 3D structure, but in this section we discuss only those which have). There are essentially two main problems. The first, what makes the protein structure stable? The second, how does the protein fold so rapidly?

Once X-rays were applied to protein crystals, it was established that they have well-defined structures. It was also concluded that in order to *function*, the protein must attain a very precise 3D structure. For instance, an enzyme would work properly only when it is in its proper 3D structure. Some experiments have shown that if one takes an enzyme solution and heat or add some large quantity of some chemicals to it, the enzymatic activity is diminished, or even destroyed. This process was called *denaturation*, meaning that the natural structure of the protein is lost. It was also known that these denatured

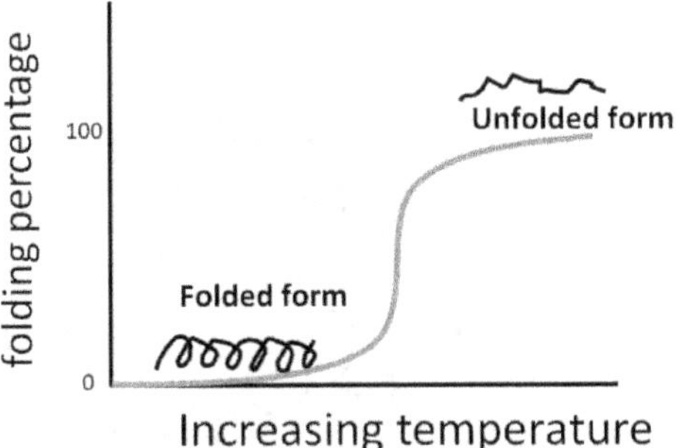

Fig. 4.33. Schematic description of the process of protein denaturation.

proteins are "useless" because they lose their ability to function. Hence, scientists concluded that in order to function properly, the protein must have a precise 3D structure. A classical experiment is described in Fig. 4.33. Take an enzyme, measure its activity, say enzymatic activity, raise the temperature and then observe the activity sharply dropping to zero (the activity in Fig. 4.33 is inversely proportional to the protein folding percentage). A common illustration of protein denaturation is the process of cooking an egg. As the temperature increases, the protein albumin in the egg undergoes denaturation and loss of solubility.

At this point Anfinsen and co-workers (1961) discovered a remarkable phenomenon. They took a protein, specifically, Ribonuclease-A (see Fig. 4.34), in a controlled laboratory condition, i.e., in a well-defined solution *in vitro*, which means in a glass, or laboratory test tube, and not *in vivo*, which means in the real environment of the cell. They found that when the solution was gradually heated, or when certain solutes were added, the enzymatic activity of the protein gradually diminished until it was lost completely. This was an expected result, as the denaturation phenomenon had already been known previously. What was remarkable in the new experiment was that if one restored

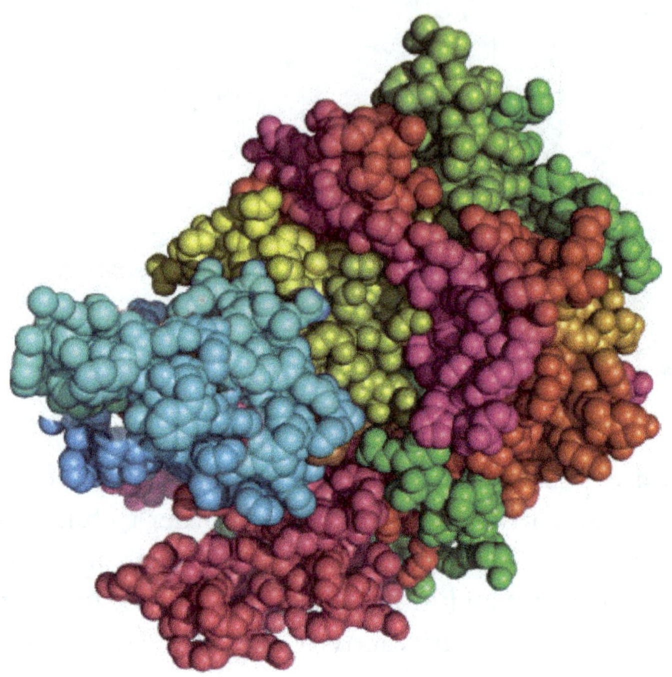

Fig. 4.34.   Ribonuclease-A.

the original condition of the temperature and the composition of the solution, the enzymatic activity of the problem could be restored.

The immediate conclusion was that the process of denaturation is *reversible*. This means that the folding of the protein into its active or native structure can be reversed, and this occurs without any agent. This conclusion could not be drawn had the experiment been carried out on the protein in the cell, or *in vivo*, since there are too many factors that can affect the newly born protein. From this experiment Anfinsen (1973) concluded that all the *information* necessary for the folding into the precise 3D structure is already contained in the sequence of amino acids. Subsequently, people have often reached other conclusions from Anfinsen's results. First, the 3D structure is essential for the activity of the protein. Second, the particular composition of the solution in the experiment is essential, and as you know the most important component in this solution is water!

The experiment and its conclusions left many questions unanswered. This is one example where one discovery opens the doors to a plethora of new questions.

As mentioned earlier Anfinsen concluded from the result of the re-naturation that the "information" on the folding was somehow already encoded in the sequence of amino acids. Here we see a usage of the term "information", not in its usual sense as we commonly read in books or see on television, but in a more abstract sense, as if the sequence "knows" the 3D target it has to reach or acquire in order to function.

Some even suggested that there might be a "code" that translates each sequence of amino acids into a 3D structure. This code, if it exists, is quite different from the code of transcription from DNA to RNA, or from RNA to protein. No one has shown that a code from a sequence of amino acids to a 3D structure exists, and we shall not discuss this question here. Instead we shall focus on the two problems mentioned above. For more details on the PFP, see Ben-Naim (2011).

The first problem that arose form Anfinsen's results is how and why proteins fold. The PFP was designated by the Journal *Science* in 2005 as one of the "unknowns of science".

"Can we predict how proteins will fold? Out of a near infinitude of possible ways to fold, protein picks one in just tens of microseconds. The same task takes 30 years of computer time."

One can see from this quotation that there are essentially two, quite different problems that comprise the protein-folding problem. The first is "can we predict how protein will fold?" The answer to this question depends on what one means by the word "predict". One can think of at least three possible definitions of this word.

First, we give you a sequence of amino acids and then you synthesize the protein and do an experiment in the lab. If it folds then you determine its 3D structure.

Second, we give you a sequence of amino acids and then you do a simulated experiment on a computer. Nowadays, people refer to an "experiment" carried out in a computer as an *in silico* experiment. This is indeed a new concept. In any case, if the protein folds, you get a 3D structure.

Third, we give you a sequence and then you read it. You do not do any experiment, but just "predict" the structure from the sequence.

Most biochemists will accept the second definition of "predict", while a few will want to have an answer of the third kind.

The second problem has to do with the speed of the folding process.

An immense effort has been undertaken by many scientists to find a solution to this aspect of the protein-folding problem. Cyrus Levinthal enunciated one of the clearest formulations of this problem in 1973. To appreciate the enormity of the problem, consider a string of about 100 beads, which represents a relatively small protein. The beads are connected by little strings, which we call a chemical bond. The whole molecule can now rotate about each of these bonds.

Look at a string of beads threaded through small sticks and start to rotate them about the various sticks. The simplest example is the case of four beads connected by three springs (see Fig. 4.35). Let us denote the beads by the numbers 1, 2, 3 and 4, and the springs by the pairs $(1, 2)$, $(2, 3)$ and $(3, 4)$. If I rotate the molecule about the axis $(1, 2)$, the shape of the molecule does not change. The same is true for rotations about the axis $(3, 4)$. You can see that all the distances between the beads and all the angles between the bonds are unchanged. However, if we rotate the molecule about the axis $(2, 3)$ new shapes are obtained.

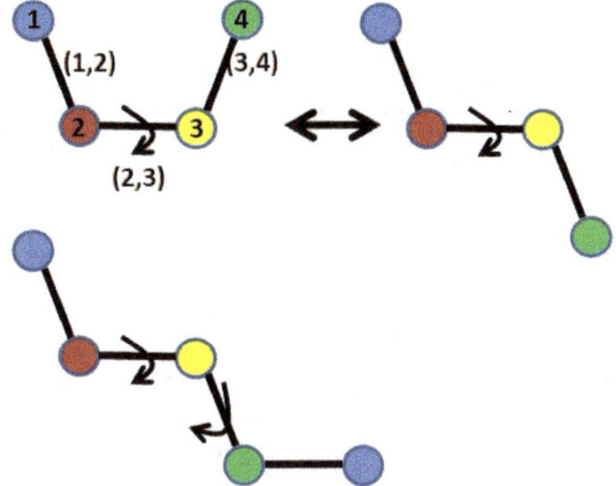

Fig. 4.35. Rotation about one and two bonds.

We call these shapes configurations or conformations. As you can see, the distances between beads 1 and 4 are altered.

If you play with a string with six beads you get many more configurations, and of course in a string of 100 or 200 amino acids there are numerous possible shapes that the protein can attain.

Levinthal contemplated the following question: "Suppose that the protein rotates about all its bonds at random, how long will it take to get to the correct shape of the native 3D structure?" What does it mean that a protein changes randomly?

Imagine a drunken man who has just left the bar after downing a bottle of wine. Can you tell how long it will take him to reach his house? Of course you cannot, because the drunken man will be walking *randomly*, meaning he will turn unpredictably, probably not even knowing which is right or left. Clearly, if the city is very large, say like the whole of New York, it may take years and even thousands of years until he hits the target point, which we denote as Y in Fig. 4.36.

Levinthal made a quick estimate as to how long it will take a protein to fold to a specific 3D structure by moving randomly

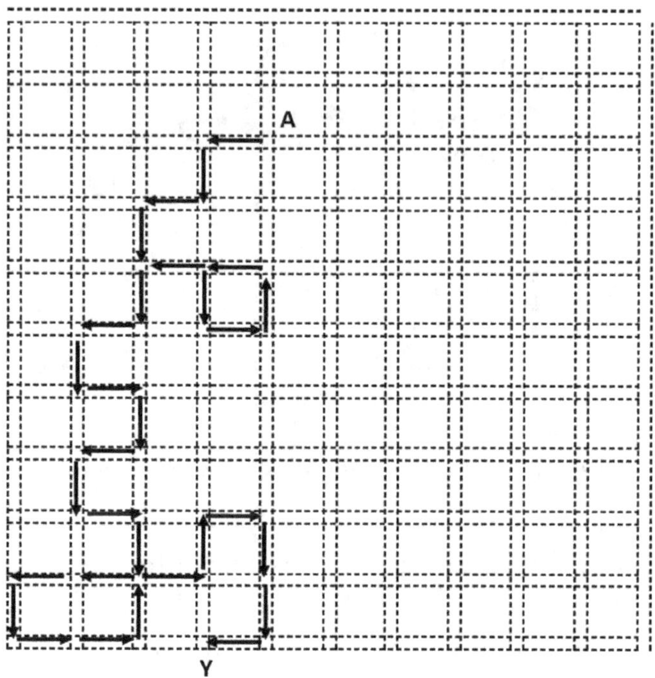

Fig. 4.36.   Walking at random from point A to point Y.

in the "configuration space". He argued that a random search in the configuration space for a 100-amino acid protein would require 10^{27} years! Too much time...

As of today, molecular dynamics simulations are giving invaluable hints on the first stages of the folding process. It is known now that the unfolded state still retains key long-range interactions and that the local propensity of the sequence to fold in a given secondary structure element narrows the "search" in the so-called conformational space. This seems to mean that biological proteins somehow evolved to properly fold. In fact, many random amino acid sequences only acquire ill-defined structures (molten globules) or no structure at all.

There are some general rules, however. Hydrophobic amino acids tend to be kept inside the structure, with little or no contact with the surrounding water; conversely, polar or charged amino acids will often be exposed to a solvent. Very long proteins will often fold in various distinct modules, instead of in a single large structure.

According to Levinthal's estimate, if the protein "walks" at random, then it would take millions or even billions of years to reach the native structure. On the other hand, we know that the protein does reach its native structure in a very short time of seconds or minutes. This seeming conflict between Levinthal's estimates and the experimental fact is known as the Levinthal paradox. A great effort was expanded to solve the so-called Levinthal's paradox. Before we explain the solution to this paradox, we note that many phenomena in science were explained either by "cause" or by "purpose". To understand the difference between these two approaches, let us go back to the analogy of the drunken man walking in a big city. Suppose the drunken man is walking in the streets of New York and randomly turns at each intersection, either left or right or perhaps even going forward and backward. How long will it take for him to reach a specific building, say the Empire State Building?

Now we go back to the subject of proteins and Levinthal's question. Levinthal had wondered how it was possible that the denatured protein reached the same structure every time, i.e., starting from A, the denatured state, and reaching Y, the native structure. He estimated that if the motion was really random it would take eons to reach point Y, but in reality it takes only a few minutes. This situation was considered a paradox. Levinthal did not consider this a paradox. He recognized that the "motion" of the protein was not random, or at least not totally random. Levinthal did not know what it was that made the protein "walk" in some preferred pathway. This was the essence of the protein-folding problem that many scientists struggled to resolve for over 50 years.

You probably might have heard the story of someone who lost his wallet while walking at night. He could not see the pavement and so he went to the nearest street lamp and searched for his wallet under the lamp.

Well, the situation in protein-folding is perhaps funnier. What if the wallet really fell under the lamp but the man did not see it and went on to look for it in the dark, where he obviously could not find it?

This happens quite often in science. You search, not for a wallet, but rather for a solution to the problem. You believe that it is somewhere and you try desperately to find it where it cannot be found. Similar situations occurred in the case of protein-folding. Many scientists searched in the wrong place.

Some thought that something *pulls* the protein towards the target (point Y). Some suggested that there might be a *code* that translates from point A to point Y, from a sequence to structure. Others suggested that something *pushes* the protein towards the native structure, or towards point Y in our example. But no one figured out what the mysterious agent that pushes or pulls is. A different school of thought is that Evolution has created a "code" telling the protein how to get from point A to Y, and all we need is to decipher that "code".

It should be said that even in the Evolution theory, there is no target. Evolution was never presented with the problem of protein-folding. Thus, it never solved that problem. Instead, during Evolution some sequences were synthesized at random. Some folded, while others did not. Of those that folded, some were stable and some were not. Of those that were stable, some had evolutionary advantages, and therefore they survived. It is the result of the long journey of evolution that we now know which protein had survived. One cannot ask: How does a protein "know" how to fold? It does not know of course. It was selected by evolution because it functions.

As in the example of the drunken man, if we are not aware of any force that pushes the man along a pattern of directions, we might suspect that the man somehow "knows" the target he has to reach. But the protein molecule does not know anything. If we observe that in every experiment, the molecule reaches the same target, then we must conclude that there is some strong *force* that *forces* the protein to "walk" along a small range of specific pathways leading to the target. Thus, what appears to us as if the molecules "know" where they are going is only an illusion. The fact is that the protein is coerced to walk in some specific preferential patterns of directions leading to the target. Look again at the pattern of the winds in Fig. 4.37.

Now, choose any point in the city and put the drunken man there. You will see that every time he starts walking from any

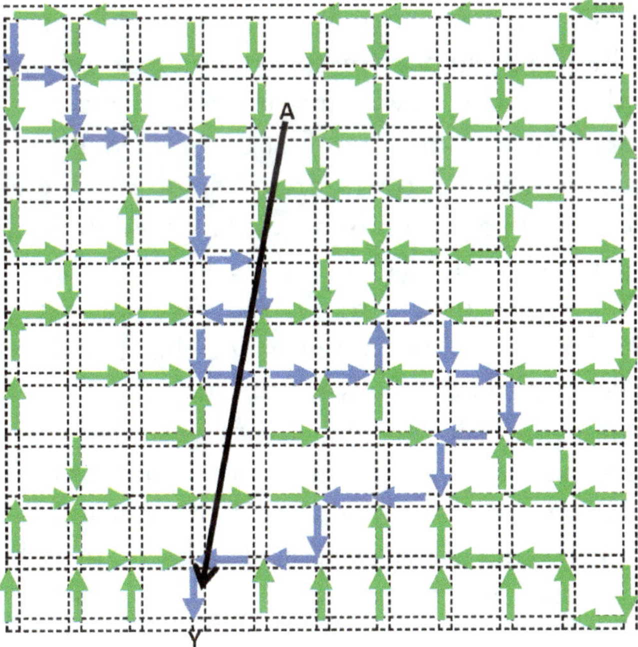

Fig. 4.37. Walking at random in streets where there are strong winds along the green arrows.

initial point A, he will end up at point Y. This is of course the result of my *design* of the pattern of winds. But Nature did not *design* any pattern of forces. This pattern of forces has evolved during a very long period of time.

4.4.3. *What is the Role of Water in Protein-Folding?*

Today, almost everyone agrees that water plays a central role in maintaining the stability of proteins. In the early 20th century people believed that hydrogen-bonding was the most important factor. Indeed, Linus Pauling argued that hydrogen-bonding was fundamental in maintaining the stability of proteins.

Regarding the stability of proteins, there are two schools of thoughts or doctrines — one maintaining hydrophobic, one hydrophilic. These are very different, yet they have a common element which is that water is important. The same goes for speed — there are two doctrines, very different but they have in common the accepted view that water is essential.

In 1959, Walter Kauzmann first enunciated a brilliant idea. Kauzmann knew that proteins are made of 20 different amino acids. Each amino acid can be viewed as a solute in the experiment described earlier. That means we can classify all the 20 amino acids according to their hydrophobicity scale (see Fig. 4.38). Thus, we have amino acids that are hydrophobic, hydrophilic, and some in between the two.

Kauzmann noticed that when proteins fold, most of the hydrophobic amino acids are found in the interior of the folded protein, while most of the hydrophilic amino acids are exposed to the solvent. This was an important observation. It reminds us of the distinction between the two kinds of solutes.

Kauzmann made the following connection between the two phenomena; the distribution of solutes (in our case amino acids) between water and oil, and the distribution of the same amino

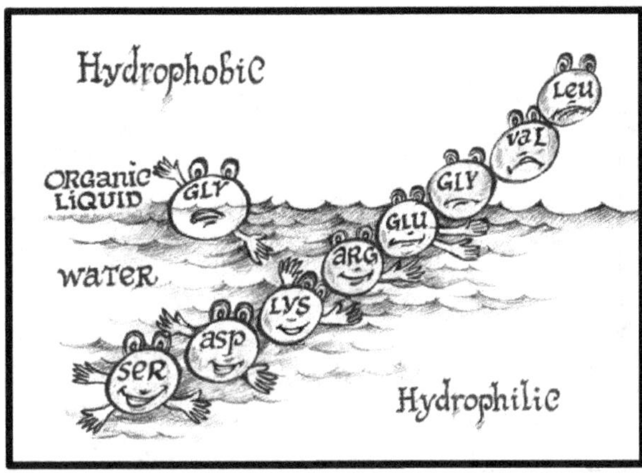

Fig. 4.38.   Schematic description of the hydrophobicity scale.

acid, but now not as a free solute but as one "bead" in a long chain of amino acids. He postulated that whatever the driving force that forces hydrophobic groups to prefer oil over water is, the same driving force will apply to each hydrophobic amino acid to prefer the interior of the protein rather than being exposed to water. Here, the interior of the protein takes the role of the oily phase. The two experiments are shown side by side in Fig. 4.39.

It should be stressed, however, that Kauzmann made mental connections between the two experimental facts. There is nothing in this comparison that refers to any microscopic properties of the amino acid. In fact, Kauzmann used the concept of hydrophobic "bond" to refer to the tendency of the hydrophobic groups to prefer the interior of the protein. The term hydrophobic bond was soon replaced by the hydrophobic effect, simply because there was no real bond here, and the term hydrophobic bond might be misleading.

In fact, not only was there no bond, there was no molecular interpretation of the hydrophobic effect either. Nevertheless, the concept of the hydrophobic effect had gained an

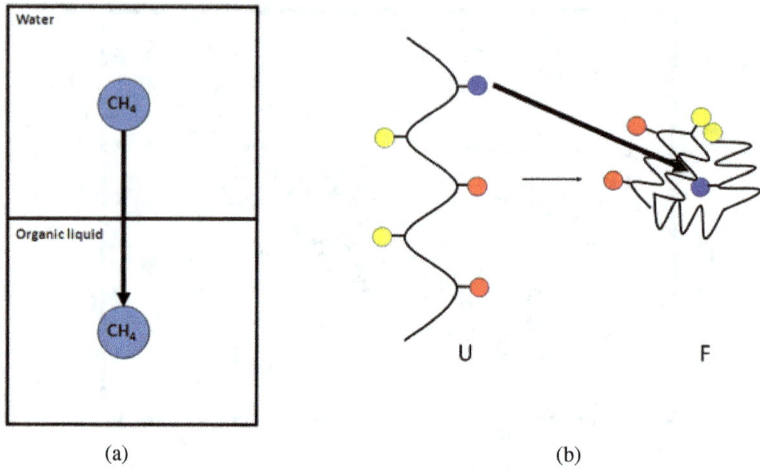

Fig. 4.39. Kauzmann model for the hydrophobic effect.

almost universal acceptance. The protein-folding was explained by the hydrophobic effect. The association between proteins was also explained by the hydrophobic effect. Unfortunately, the hydrophobic effect itself remained unexplained on a molecular level.

For many years, scientists were so impressed with the hydrophobic effect argument that they did not look for any alternative explanation. People imagined that the protein as a whole "collapses" in such a way that all, or nearly all the hydrophobic groups were removed from water. The protein could be viewed as a micelle, but made up of a single molecule. A micelle is an aggregate of many molecules, each possessing two parts — a "hydrophobic" and a "hydrophilic" part. What water does in effect is remove the hydrophobic parts from the water while keeping the hydrophilic parts in the water.

Remember that hydrophobic molecules prefer to be in the organic liquid and therefore will move from water into the aqueous environments. In the same way the hydrophobic "beads" along the protein are in the interior of the protein and are not exposed to water. The same is true in the process of micelle formation.

Although Kauzmann's idea was very plausible, it was never proven correct. The amino acids, being part of the chain of amino acids behave quite differently from free amino acids. It is true that most of the hydrophobic amino acids are found in the interior of the protein in its Native structure. However, this in itself does not provide an explanation for the folding process.

A detailed examination of all possible solvent-induced effect shows that not only was Kauzmann's model inadequate, it also revealed a new and rich repertoire of effects involving *hydrophilic* amino acids. For our purpose, these may be viewed as tiny arms protruding from the proteins and reaching into the water. These arms can form bonds with water molecules.

The moral of this story is that we should not rely only on what we see. One famous example is the theory that the earth is at the center of the universe. We see the sun rising in the east, then going over our heads, finaly setting in the west. What can be more convincing than the conclusion that the sun rotates around the earth? In fact, people believed for many years that the sun, as well as the stars in the sky rotated around the earth.

People were puzzled as to how proteins find their way to the final active structure so quickly.

We should mention again that many proteins fold with the help of some agents called chaperons or chaperonins. It should be mentioned that these molecules help the protein to fold. We shall be interested in the folding process in aqueous media. It is said that proteins fold spontaneously. Well, it is not exactly spontaneous — they need the help of water molecules.

In an *in vitro* experiment conducted under well-controlled conditions in the laboratory, Anfinsen discovered that protein denatured when the temperature increased or when he added some chemicals such as urea or guanidine. Now the nice thing about his experiment was the following: In the denatured state the protein did not function. For instance, if the protein is an enzyme one can check the enzymatic activity in the denatured state and find none or almost no activity. Now the surprising thing was when he restored the original conditions, the protein regained its enzymatic activity, which meant that the specific 3D structure of the protein was restored.

Note carefully that we do not know exactly in what form the protein is ejected from the production machine. We only know that after a short period of time the protein acquires its specific 3D structure, so that it becomes functional or active. You can imagine that the ribosomes produce a long string of beads. As such it can do nothing. It is "useless". But after a short period of time, it folds on itself and is activated as if becoming a living robot. Perhaps you have heard about the story of Golem, which was made out of clay but came to life. We also do not know exactly what the steps are through which the protein folds *in vivo*, whether the denaturation–renaturation of the protein *in vitro* is identical with what actually occurs in the cell. What we are striving to understand is how the protein in the *in vitro* experiment folds so rapidly into its final structure. It is clear that water molecules play a major role.

Remember what we have learned about hydrophobic and hydrophilic molecules? As we have seen, the hydrophobic solutes have no "arms" and they are not "welcomed" by water molecules. On the other end, hydrophilic "water loving" solutes have one or more arms and they can hold hands with water molecules.

Now the protein can be viewed as a string of solutes, some of them having no arms, while others have. In reality, there are many more arms in the protein than the number of amino acids.

Thus, the traditional explanation of protein-folding is that all hydrophobic amino acids tend to be expelled from the water and therefore the protein folds into a kind of compactly packed molecule, much like a micelle. You remember that micelles are made of many molecules, each having a hydrophilic and hydrophobic part. The protein is a giant molecule that has both hydrophobic and hydrophilic beads, so the folding is explained as a result of the tendency of all the hydrophobic beads being expelled from the water. The other view is quite different. The water molecules can do the folding by pulling the arms of the *hydrophilic* beads. One water molecule can do this on a very small segment in the protein, a region where the two arms that belong to the protein happen to be at such a distance that a water molecule can grab one of them and bring the other closer to the first.

This is how α helices and β sheets are formed. These are two periodic structures, but the result of these grabbing and pulling does not need to lead to a regular periodic structure.

Can you imagine that a protein of say 100 amino acids has over 200 of such arms that are surrounded by water molecules, and that each of these water molecules can do the same small feat but together they can bring the apparent random configuration of the protein into its final structure, the native structure?

You should also realize that at each step of the folding the water molecules do not necessarily lead the protein towards the target. At some stages the protein may move away from the "target". However, the concerted effort of so many water molecules simultaneously grabbing and pulling the arms on the protein eventually leads to the final 3D structure.

So why is the protein-folding considered a big problem?

The reason for this lingering mystery is that many scientists wasted much time and effort in searching for the answer to Levinthal's question in the wrong direction. However, you should know that for a long time people did not pay much attention to the role of water in the process of protein folding.

Even after it was recognized that water is vital for the process of protein-folding it was not exactly clear how water does that job.

Remember the metaphor of the drunken person walking randomly in a big city? Well, many scientists thought the protein-folding process is a target-directed process, as if there was some "code" that commanded the protein to move *towards* the target; the target being the final 3D structure of the native protein. Very little effort was expended in identifying the *forces* that acted on the protein. Furthermore, those who did search for the causes missed the correct cause.

4.5. The Role of Water in the Self-Assembly of Proteins

Atoms like to be free to wander around and visit all the points in the volume they occupy. The higher the temperature, the larger the speed of the flight in between the collisions is. However, if a strong chemical bond binds the two atoms together, the pair of atoms loses some of their individual freedom. They can still wander about in the space, but now as a pair and not as two separate individuals.

At each temperature there is a competition between two tendencies — one is the tendency to have individual freedom and the second is the strength of the chemical bond that forces them to stay together. This is referred to in technical terms as the competition between entropy and energy. The entropy is represented here by the tendency to be free, and the energy is represented by the strength of the bond, hence the tendency to stay "in touch". As we increase the temperature the tendency towards freedom increases and the rate of dissociation, or if you prefer to call it, the rate of the "divorce" increases.

These processes are what we actually observe in any chemical reaction involving association. These could be two hydrogen atoms forming a hydrogen molecule, two hydrogen atoms and one oxygen atom to form a water molecule, three hydrogen atoms and one nitrogen atom to form ammonia and so on. In all of these chemical reactions the final state is a result of the balance between the two competing tendencies — the entropy and the energy — or in simple language, between individual freedom and the strength of the bonding or the "glue". Entropy and energy are two fundamental concepts in physics and chemistry. We do not need to discuss their precise meaning here, but just think of entropy as a measure of the tendency of individuals to wander freely and reach every corner of the box. On the other hand, the energy in our case is a measure of the strength by which the pair of individuals are attached, or if you like, "glued" to each other.

Let us briefly analyze the energy and entropy contributions for a chemical reaction. Let us focus on the formation of a hydrogen molecule according to:

$$H_{(g)} + H_{(g)} \rightarrow H_{2(g)}.$$

According to thermodynamics, the Gibbs energy change for this process at a constant temperature can be written as:

$$\Delta G = \Delta H - T\Delta S,$$

where ΔH and $-T\Delta S$ represent the energy (enthalpy) and entropy contributions, respectively.

The reaction we are considering is exothermic ($\Delta H < 0$) because the formation of a covalent bonding (H–H) gives rise to a more stable system (H_2 is more stable than $H + H$), the energy difference being released to the surroundings. But on the contrary the molecule formation gives rise to a positive entropy contribution because ΔS is negative. Two hydrogen atoms love to be together by energy considerations, but they tend to be separated by entropic arguments. The net contribution is, in this case, negative ($\Delta G = -406.5\,\text{kJ/mol}$ at $25°C$ and 1 atm) and the process is said to be spontaneous — two hydrogen atoms "feel more comfortable" when they are forming a hydrogen molecule than when they were freely wandering around.

Now you might have asked yourselves what do all of these have to do with proteins, or with biology? Here is a profound mystery, or at least, a mystery until very recently. Proteins wandering about in the solution would like to be free of any attachments; no strings attached. However, in reality we know that they form pairs, triplets, quadruplets and much larger complexes, which live for long periods of time. Unlike the two hydrogen atoms that succumb to the dictates of the powerful chemical bond that coerces them to stay together, nearly forever, there are no such bonds that "glue" the two proteins together.

Imagine two proteins in a gaseous phase.

We can assure you that if we could have two protein molecules in the gaseous state they will, for most of the time, enjoy their freedom as individuals. From time to time, they might collide and perhaps even stay together for a fraction of a second, but nothing like a permanent relationship. However, the same two proteins in aqueous solutions, at certain specified conditions would form a permanent marriage and will stay together for a long time, nearly forever. Why? This has been a great mystery for a long time, not less the mystery of the folding protein. For quite some time, it was suspected that water is essential for this binding. But what is exactly the factor that binds the two proteins? It was such a big mystery that the editors of *Science* listed this as one of the "Unknowns of Science" right after the protein-folding process. It is of interest to quote the precise formulation of the problem as stated in *Science* in 2005:

> *"How do proteins find their partners? Protein-protein interactions are at the heart of life. To understand how partners come together in precise orientations in seconds, researchers*

need to know more about the cell's biochemistry and structural organization."

There are essentially two problems: finding a partner and staying together. In the case of protein, the main question is not how the partners find each other, but finding the "glue" that keeps them attached (see Fig. 4.40). There is another problem that we shall discuss later and it has something to do with "molecular recognition".

This is not the same as molecular recognition used in daily life. The "recognition" in that phrase is not the same as what you might have guessed, so do not waste time on this concept. We shall relegate it to a later discussion. For many years, the self-association of proteins to form bigger units such as hemoglobin was quite a mystery. We shall also discuss this later. Two single proteins in a vacuum or a non-aqueous solution will not associate. They would like to be free to wander around with no strings attached. But in a pool of water they do associate.

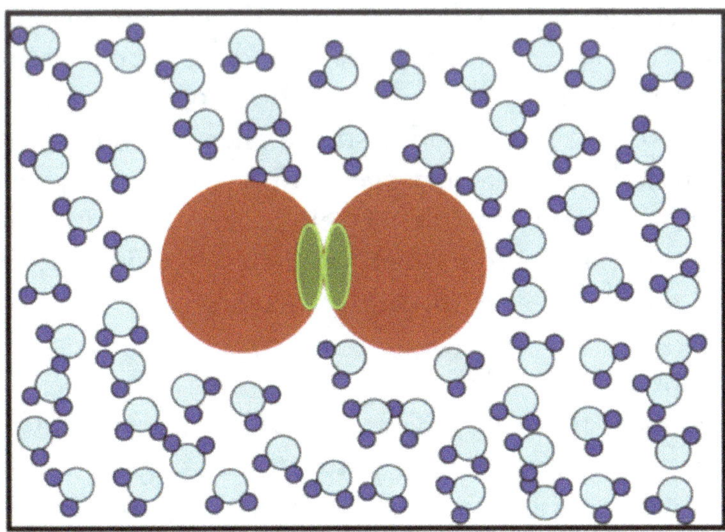

Fig. 4.40. What is the "glue" that binds the two proteins?

It looks to us as if the water serves as a "glue" to bind the proteins. Again, "glue" is used here as a metaphor.

But what factor coerces the proteins to sacrifice their freedom and form a strong, stable and long-lasting pair? You might think that there might be some chemical bond similar to the one in hydrogen molecules, but no such strong chemical bond was found. You might think that there is some matchmaker who helps the pair to find each other, and makes sure that they will also stay together, or perhaps the two proteins love each other so much that they simply love to stay together voluntarily. But why do they love each other only when they are swimming in a pool of water?

It may come as a surprise to you but the concepts of love and hate were in fact used in connection with protein-binding. They were used not in the same sense that we use them in human lives, but metaphorically.

If there are enough proteins around, they will surely collide with each other — the larger the density of the proteins, the larger the frequency of collision. Thus, the main problem is not *how* the protein finds its partners, but discovering the factors that bind them together for long periods of time. This is the mystery. No one has found a strong chemical bond between the two proteins. Unlike hydrogen atoms, which are spherically symmetric, the surface of the protein is very different at different directions. There is another problem: How do the proteins "know" which surfaces should face each other in forming the stable pair?

If there is no chemical bond between the two proteins, scientists suspect that the solvent, in particular water, which is the main component in the solvent surrounding the protein, is responsible for binding, i.e., the solvent serves as the "glue" that binds the two proteins. That conjecture is correct, but how does this glue work on a molecular level?

As in the case of protein-folding, the search for an answer to this question went astray. Kauzmann's brilliant idea about the hydrophobic effect was invoked in explaining the protein–protein association. This idea made sense. If the two proteins "hate" water, they might come close to each other, and or at least minimize the contact with water, thereby avoiding contact with water. We should also tell you an important finding that supported this conjecture. This is the story of the sickle cell anemia.

Hemoglobin is an important and among the best known protein. We shall discuss its function as an oxygen carrier later on. We want to tell you the story of an abnormal hemoglobin molecule referred to as hemoglobin S, or sickle cell hemoglobin.

This story is very exciting since this was the first time it was speculated that a disease called sickle cell anemia could result from a small change in the sequence of the proteins. More specifically, it was found that in the abnormal hemoglobin S, one amino acid — glutamine — is replaced by valine. The important aspect of this replacement is that one hydrophilic group was replaced by a hydrophobic group. It was shown that this replacement caused excessive association of the abnormal hemoglobin, leading to a deformation of the red blood cells (erythrocytes), from roughly disk-like to sickle-like form. The latter cannot flow as easily through the capillary blood vessels, thereby causing a painful, debilitating and oftentimes fatal disease that is known as sickle cell anemia.

We brought up this story to show the reader how a small change in the surface of a protein can lead to an abnormal aggregation, or if one likes, abnormal "glue". The factor in question that made the "gluing" possible is the hydrophobic group in the abnormal hemoglobin.

Thus, the cause of an abnormal association of hemoglobin supported the conjecture that the "glue" that helps bind proteins is the hydrophobic groups. If the protein molecules associate in such a way that the hydrophobic parts are in the interface of the pair of molecules, then the hydrophobic groups will be removed from water and will be buried in the interface between the two proteins, much like what we saw in the case of protein-folding where we observed that hydrophobic groups were buried in the interior of the protein.

This explanation was so appealing that most textbooks and research articles adopted it to explain both the protein-folding and protein–protein association.

We recall that "water fearing" amino acids tend to avoid the water environment, and therefore they bury themselves in the interior of the protein. But here we have a different situation. The "water fearing" amino acids are *already* buried in the interior of the globular protein, so the water is already excluded from their surroundings. Therefore, how can this hydrophobic effect be applied to the case of the association?

The analogy is not perfect but as we explained before, the adherence to the dogma that the hydrophobic effect could explain protein-folding is the same reason for the hindrance in finding a solution to the question of protein association. In reality not all "water fearing" or hydrophobic groups find themselves buried in the protein's interior. Some are still exposed to water. Thus, in the globular protein, you can find some stretches of surface areas that consist of hydrophobic groups.

When the two globular proteins are separated the hydrophilic groups on their surfaces are happy to be exposed to water. On the other hand, the hydrophobic groups are not happy to be in contact with water. These groups are similar to the hydrophobic groups in the unfolded protein that went into the interior of the protein. But now the hydrophobic groups that were left behind are frustrated since they are still being exposed to water.

What do you think they will do? Well, they will do exactly the same as in the folding case. If two proteins can associate in such a way that the hydrophobic stretches on the surface come together, they will get rid of the surrounding water molecules. In forming the pair, the analog of the interior of the protein is now the interface between the two proteins. In both cases the hydrophobic groups that were initially exposed to the water, get away from water and feel more comfortable amidst their "kin".

As in the case of protein-folding this explanation was quite appealing and very convincing. In both cases, the "water fearing" groups got rid of the surrounding water molecules. Unfortunately, calculations have shown that while this removal from the water was indeed a real phenomenon, this effect was not strong enough to explain why any two proteins would sacrifice their freedom without knowing what the trade-off was. In other words, when two proteins bind together some of the hydrophobic groups are indeed in the interface between them, and therefore they achieved their aspiration to do away with contact with water, but it was not clear at all that this factor was sufficiently strong to hold the two proteins together for a long time.

It turns out that the "water loving" or hydrophilic groups on the surface of the protein, rather than the hydrophobic groups, are the ones that drive the binding of the protein and pay for the loss of freedom of the single protein.

People believed that the tendency of the "water fearing" amino acid to avoid water is responsible for both protein-folding and protein association. Both of these phenomena could not be explained satisfactorily in terms of the hydrophobic effects. On the other hand, the hydrophilic effects came to the rescue in both cases, transforming a totally mysterious phenomenon into a clear and almost obvious one.

When water molecules form a bridge between the two proteins, it is much like a real bridge that connects two regions.

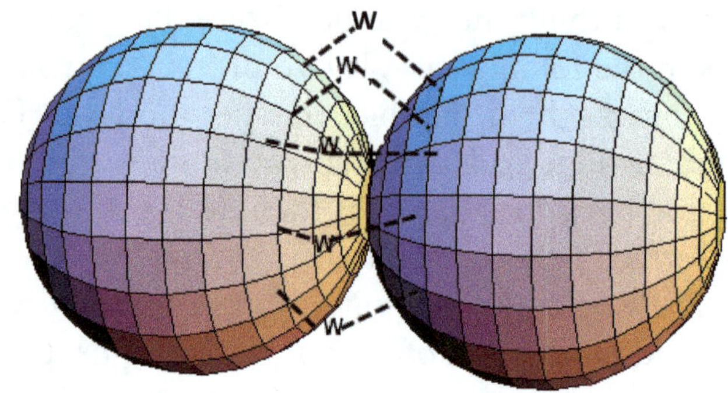

Fig. 4.41.   Two proteins "glued" by water molecule bridges.

But note that this bridge is not permanent. As you can imagine, a single water molecule might get "exhausted" holding the two proteins together for a long time. The water cannot keep holding these two huge proteins together for too long.

From Fig. 4.41 the reader can see the concerted efforts of many water molecules in bringing together two proteins in the form of a sequence of bridges (that look like zippers) that tighten the connections between the two proteins. It is almost like a pattern of stitches of water molecules connecting the two proteins.

As one can see, a single water molecule cannot do much. It can hold one, or two, or even three arms for a short period of time, but after that the proteins will slip away. On the other hand, the cooperation of many water molecules will make a big difference. Together, they can exert a powerful force that can hold the proteins together.

The same scene repeats itself on all the other sites, so statistically the net effect is that there are always enough water molecules that will do the job of bridging the two proteins. This effect is statistical, not permanent. If one looks from afar, one will see an intact sequence of stitches. In reality, if one looks closer, one will see water molecules exchanging roles continuously. Some hold two arms, while others hold only one arm but try to grab another arm. Some "get tired" and go away but are instantly replaced by eager newcomers who perform the same task. Therefore, what one sees from afar is quite different from seeing the same thing up close. We hope that by now the reader can understand in what sense the water molecules serve as "glue". It is not real glue that fills up the space in between the two proteins, but glue only in a metaphorical sense.

It should be noted that there is no direct chemical bond that holds the proteins together. This is not like the case of two hydrogen atoms forming a hydrogen molecule as a result of chemical bonding. There might be a few arms of one protein holding hands with another few arms of the second protein, but this is not very significant. It is certainly not sufficient to hold the two proteins together for any length of time. What holds the two proteins together for a long time is not by the direct *forces* between the proteins, but rather by *indirect bridges* that stitch the two proteins via the many arms that protrude from their surface. The net effect then is a result of the combined efforts of many water molecules.

The combined efforts of many water molecules have a dramatic effect on the protein. This is one reason why water is so essential to life. You have seen their role in protein folding, and now you have also witnessed their role in protein–protein association.

What we have learned so far is only a first step in the process that scientists refer to as self-assembly. You saw how two proteins associate. But the process can continue to four, six, ten, hundreds and even thousands of proteins. It gets more complicated but the "drive" for all these processes is the same. Recall that initially the concept of hydrophobic effect in protein-folding was so convincing. In the folding process, proteins do have most of their hydrophobic groups in their interior. Therefore, it is natural to think that this effect is also the driving force for the folding process. But in the association process there were not too many hydrophobic groups in the interface between the two proteins. So it is the hydrophilic groups that do the job of "gluing" together the proteins. Thus, although these hydrophilic groups "love" water molecules, they help to bring the two proteins together, as if to remove them from the water environment.

4.6. Molecular Recognition

We shall discuss molecular recognition. Everyone knows what "molecular" and "recognition" mean. However, when put together "molecular recognition" is quite different from what we obtain by combining the two concepts. Molecules do not recognize other molecules in the sense that people recognize each other. Human recognition is a complex process that occurs in the brain. One sees the form or the color of a face or an object, listen to the sound of a voice, and then conclude that one has identified the person or the object.

In order to reach such a conclusion, the brain processes all the "information" it gets from the senses and compares it with the information that is stored in it, and then makes the decision; recognized or not.

Nothing like this goes on in the "brain" of the molecules when it makes the decision we call "molecular recognition". Nevertheless, the concept of molecular recognition is extremely useful and scientists use it in connection with almost any process of binding two macromolecules, or binding a small molecule to a protein or to DNA.

Basically, what we mean by molecular recognition is very simple. Suppose the surface of the protein looks like this in Fig. 4.42.

A small molecule of a certain shape (see Fig. 4.42) approaches the protein. It can hit the protein at many points

Fig. 4.42.   Binding according to the lock and key model.

Fig. 4.43.   Lock and key.

on its surface. What will happen when it hits that part of the surface that exactly fits the shape of the small molecule?

Clearly, the small molecule will bind more tightly to the site in which it fits best. Similarly, there are many ways in which the two proteins can bind, but there is only one binding mode that is preferable.

Clearly, the molecules do not "recognize" each other in the sense that human beings do. However, what we mean by molecular recognition is that the molecules bind at a specific location *as if* they "recognize" the best, or the most suitable location.

This model of recognition is known as the "lock and key" model. If you have a key, can it open any lock? See Fig. 4.43.

Hermann Emil Fischer, who won the 1902 Nobel Prize in Chemistry, proposed the "Lock and Key" model to visualize the interaction between a substrate and an enzyme.

Only the key that fits the lock will open it. Similarly, only the molecules that fit as perfectly as possible will bind. The reason is that when the two surfaces fit, there is a maximum interaction between the two surfaces, which means a more stable bond.

As we mentioned earlier, there are actually two phenomena that calls for an explanation: (1) What is the factor that makes the two proteins bind and stay together? (2) What determines the specific mode of binding?

For a long time, the metaphor of the "lock and key" was used to explain the binding of small molecules, like drugs to protein, protein to protein, or even protein to DNA.

We have seen how a small molecule fits tightly to a specific location on the surface of the protein. Similarly, two proteins can bind by fitting the shape of one surface (the key) to the shape of the second surface (the lock).

However, when water acts as "glue", the story is quite different. Look at the following two proteins. Figure 4.44(a) shows the binding by the "lock and key" model by means of hydrogen bonds between the solute and the water molecules. You can see that the ligand L "fits" better to site A. However, when there is a possibility of hydrogen bond formation between groups on the two solutes and a water molecule, then the ligand L will preferentially bind to site B.

Here as in the case of protein–protein association, you can see that water not only provides the "glue", i.e., the force that binds the two proteins, but also selects the precise surfaces of the two proteins that will force each other in the bounded pair. In reality, the binding can be affected both by the shape (lock and key), as well as by the "gluing" water, in what was previously referred to as "water stitching" together the two proteins.

Now that you understand what is meant by molecular recognition, let us proceed with some examples where the

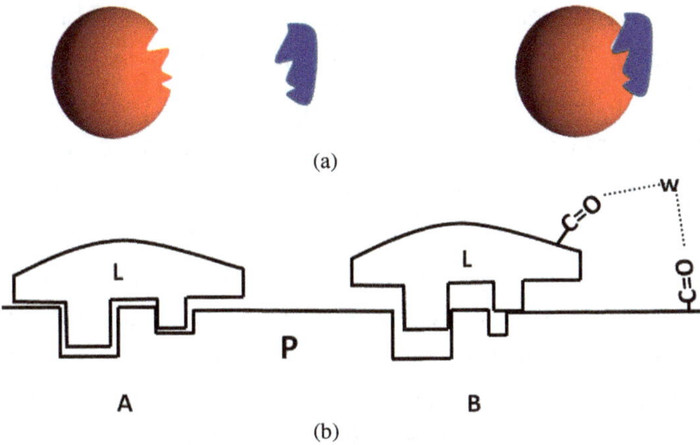

Fig. 4.44.   (a) Lock and key model, and (b) preferential binding to site B due to hydrogen-bond-bridge by a water molecule.

action or the function of the protein crucially depends on such recognition.

The simplest example is the binding of small molecules, referred to as substrate, to the active site of the enzyme, similarly, the binding of a regulator on the regulating site on a regulatory enzyme. Without molecular recognition the enzymes would not act specifically on a specific substrate.

There are many other cases where molecular recognition is vital for the function of proteins. Perhaps, the best example is the immune system. Here, scientists not only apply the concept of molecular recognition, but also the concept of "memory".

Our bodies are constantly exposed to dangerous invaders. These could be toxic molecules like snake venom, or viruses and bacteria. The first line of defense is our skin, which is impenetrable to many substances. However, many insidious substances can penetrate our lungs or intestines. For these types of invaders, the body needs a whole army of soldiers who will identify the harmful invaders and destroy them. The immune system distinguishes between friend and foe, or between self and non-self. The whole idea of vaccination is based on our knowledge of how the natural immune system works.

At the heart of the immune system is the molecule, which is referred to as an antibody. This is a complex protein that can identify foreign molecules. A typical antibody has a Y shape, and the base of the Y letter has a constant structure while the two "arms" of the Y are variable. It is this part that is responsible in identifying, binding and destroying any foreign invader.

Once the body is exposed to some bacteria, the immune system produces antibodies specific to the bacteria or any other pathogen, which is a disease-causing agent.

The second time the body is exposed to the same bacteria it is already prepared with the specific army or antibodies to fight back and ward off the disease. It is in this sense that we use the term "memory", as if the immune system "remembers" a past foe. The vaccination is essentially "teaching" the immune system of a possible harmful invader. By exposing the immune system to a partially weakened form of a bacteria, the system identifies it, prepares an arsenal of antibodies to fight back any future attacks by the real, fully active antibodies.

It is unfortunate that in some cases the immune system fails and this could spell disaster to the whole defense system. These are referred to as autoimmune diseases.

As an example we consider the HIV virus. It was discovered that the HIV virus disrupted the immune system itself. Therefore, it was a great challenge for scientists to find a vaccine for this disease, because vaccination is akin to "teaching" the immune system. However, if the immune system fails then who is going to be taught? Therefore, alternative methods to fight back must be developed.

Another important case of regulation where the "molecular recognition" becomes important is the case for the "expression" of specific genes in specific cells. First, we should remember that a gene is a small segment in the DNA that can be translated to a specific protein. The DNA contains the information for all the proteins that are synthesized by the body. However, not all the proteins are needed in each cell. Some cells need one group of proteins; others might need another group. A cell does not expend time and effort to read, transcribe and translate genes that contain the information on proteins it does not need. Instead, it reads, transcribes and translates only those genes that produce the required proteins for that cell.

The job of selecting, activating or blocking genes is done by proteins, which are called repressors. These proteins bind to a specific point near a gene that the cell does not want to read. The mechanism of binding and then blocking the expression of the gene depends on the ability of the repressor-protein to recognize the right genes it wants to block. For this reason, this system has been called a "genetic switch", that is, the repressor can determine which genes will be switched on or off in a given cell.

Molecular recognition is also important in the construction of supra-molecular assemblies of proteins. We will mention only a few examples of such multi-subunits assemblies.

Viruses are nothing but a file of information in the form of DNA or RNA wrapped in a package of proteins. One of the most studied plant viruses is the tobacco mosaic virus (TMV), which infects the leaves of plants. The single strand of RNA is coated by over 2000 identical protein subunits (see Fig. 4.45).

It should be noted that viruses are not considered living organisms mainly because their reproduction depends on them using the machinery of a host-cell into which they penetrate. However, they do reproduce and mutate as like any other cell and in this sense they have the characteristics of a living system.

The main engine in which the information on the RNA is translated into protein is also an assembly of RNA and proteins.

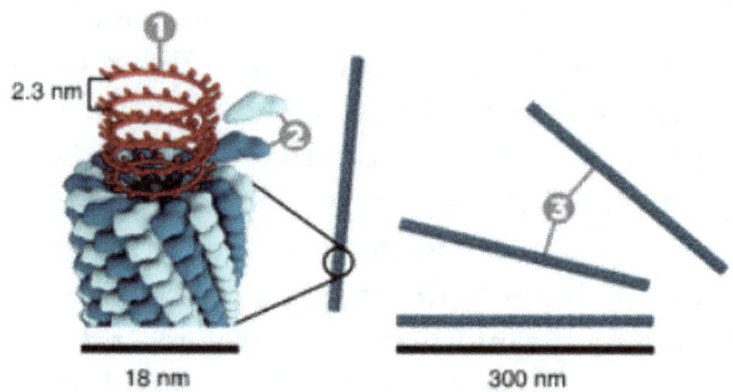

Fig. 4.45.   Schematic model of TMV virus: RNA (1), subunits of protein shell (protomer) (2), and protein shell (3).

These are the ribosomes that consist of RNA (or rRNA) and over 80 subunits of protein. It is in these ribosomes that the information carried out by the tRNA is translated, phrase-by-phrase into amino acids to produce the protein.

4.7. Protein Solubility

Most of the proteins function in an aqueous media, so in that sense it is important that the protein will be soluble in water.

Most books on proteins do not mention the solubility of proteins as a problem. If one were to ask a protein chemist: "Have you ever asked yourself what makes the protein soluble in water?" Most will immediately answer that there is no problem; the proteins are built in such a way that most of the "water fearing" groups are buried in the interior, and the surface of the protein has an abundance of "water loving" groups, meaning that these groups extend many *arms* into the solvent. Therefore, the water molecules are happy to hold hands with these arms, and that is what makes the protein soluble. It is much like a micelle wherein the hydrophobic groups are inside, while the hydrophilic groups are outside (see Fig. 4.46).

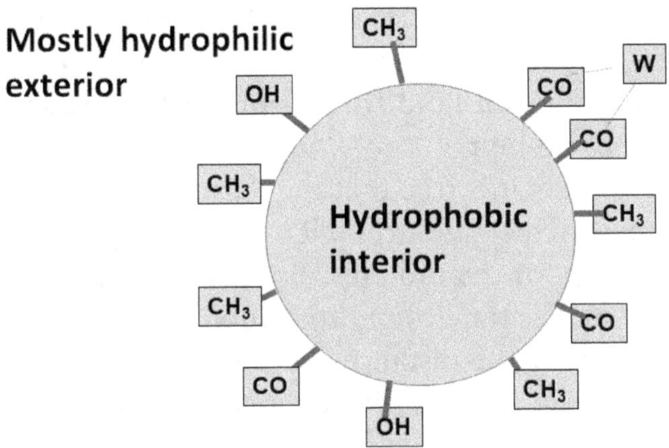

Fig. 4.46. Schematic structure of a globular protein. Most of the hydrophobic groups are in the interior while the exterior is mostly hydrophilic.

This is basically a correct answer. However, we need to add one more factor that enhances the solubility of the protein. This is a factor that not too many scientists recognize.

We consider a spherical globular protein. There are about 20 arms extending from the surface of the protein into the water, and each of these arms can hold an arm of a water molecule. As we have already discussed, this holding arms with the water molecules make the protein more soluble. Calculations show that if the arms are far from each other, so that at any given point in time an arm of the protein can form a hydrogen bond with one water molecule, then the solubility of the protein will not be large.

However, when we change the distance between some of the arms, so that the distance between two arms of the protein can be bridged by one water molecule, then the solubility will be much larger. This should remind you of the role of water in protein-folding and in protein–protein association.

It turns out that each water bridge, such as the one shown in Fig. 4.46 can cause an increase in the solubility by a factor of 100. You can imagine that in a real protein, water molecules can bridge many arms. Actually there is also a possibility that a water molecule can hold three arms of the protein, further increasing the solubility.

To summarize, it is true that the "arms" of the amino acids of the proteins, which are exposed to the water, are responsible for the solubility of the proteins. But it is not only the *number* of arms that makes the protein so soluble. More importantly is the distribution of the distances between the arms that makes the water bridges possible, hence increasing dramatically the solubility of the proteins.

CHAPTER 5

Water Resources and Their Vital Role in Survival of Civilizations — Then and Now

Water is essential not only to the very needs of life, but also to the expansion of civilization. Wherever they are, people need water to survive. The history of mankind is interwoven with the decisive role water has played in every aspect of human life on Earth. The abundance of water resources determined the fates of communities, states and empires.

The utilization of water resources had a pivotal role in the development, persistence and ultimate downfall of the world's greatest civilizations. History tells us that the societies that took advantage of their water resources for expanding economic capacity, industry and domestic consumption are those that flourished, while those that were unable to do so faced unavoidable decline.

Water has been the main source of energy in past centuries, and not many of us are aware of the vital role of water in technologies of the past. As a matter of fact, water is also needed as an energy source today in hydraulic power stations.

Hydraulic power has been analyzed in a Chapter 3. The principle is simple: One exerts a very small force and what a miracle an immense force depending on the structure of the machine is created (see Fig. 3.27 in Chapter 3).

The relation of water to technology in the history of mankind is twofold: (1) Technology was constantly improved in order

Fig. 5.1.   Steam engine. The water wheel first served as a main power source of water propelled by rivers and waterfalls. Then the steam engine arrived in the 17th century.

to handle flooding rivers, create and use canals and aqueducts, build dams and develop means for water withdrawal and distribution, and (2) water by itself has been an essential mean of producing power throughout history up to modern time, first in the form of the water wheel propelled by strong streams or falls, and later came the invention of steam engines leading to the Industrial Revolution.

Improved navigation and ship design also contributed to Europe's rise. Utilizing water resources in its eastern half, the United States (US) made proficient use of the water turbine in factories while the nation as a whole exploited hydropower to provide cheap and accessible electricity. The US demonstrated again that the civilizations that best harness their water resources and keep innovating their water infrastructure find themselves at the top of the heap.

In his book *water*, Steven Solomon offers the first-ever narrative portrait of the power struggles, personalities and breakthroughs that have shaped humanity.

Solomon presents the conception that throughout history, the most successful civilizations were those that best utilized their water resources. Historically there are five ways civilizations have used water: (1) Domestically for drinking, cooking and sanitation, (2) in economic production for agriculture, industry and mining, (3) power generation (waterwheel, steam and hydroelectricity). The Industrial Revolution came about in two phases — the first centered on textiles and on the use of the water wheel, and the second is the invention of the steam engine that helped to serve a centrally organized production model, i.e., a factory. The whole industrial world has been based on water. (4) Transportation, and (5) environmental benefit: Sustaining the vital ecosystem against man-made and natural degradation. As technological innovation and expertise progressed, access to irrigation water became easier and agricultural production increased thus leading to an upsurge in human population. The reliance on water to produce food for an ever-growing population continues today.

Another interesting point is raised in *water*: Trading among the various city-states along the shores of the Mediterranean coast developed because of limited water resources. As beautifully described by David Abulafia in his splendid book *The Great Sea*, trading encouraged the early Mediterranean states to develop early forms of market-based economies with a vested citizenry. For many of these city-states, survival was dependent on the development of a worthy navy. Rome's rise to superpower status occurred only after wrestling control of the sea-lanes of western Mediterranean. Often, Rome's most successful periods were those during aqueduct construction and an expanding water supply. There is little doubt that this ability to exploit water resources led to Rome's unparalleled rise to power.

Water has been the basis of high technology in past centuries — every period in history had its particular high

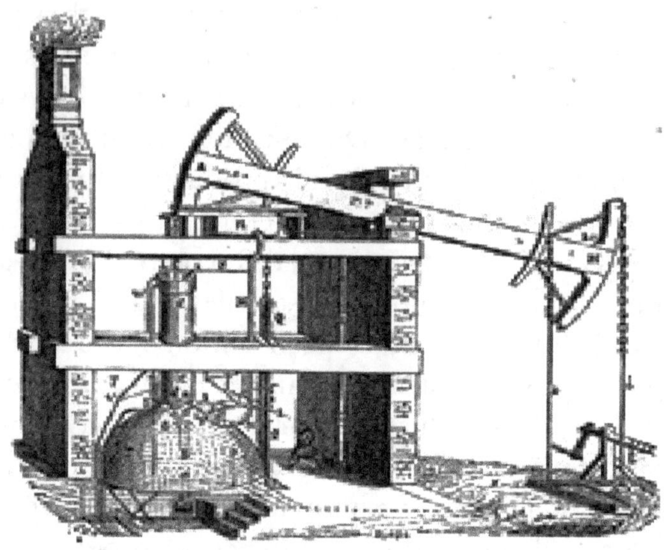

Fig. 5.2.   Engraving of Newcomen engine. based on engraving from 1717.

technology, starting from ancient Egypt (the pyramids and Nile-based irrigation systems) to the aforementioned technology in Rome.

Water supplied the power for industry prior to the discovery of electricity. Water replaced wood fuel in England in the 19th century by using gravitation force in water wheels, and then by using water in steam power. Later the steam engine arrived, serving factories to enable economic growth.

Water covers 70% of our planet, so it seems that there will always be plenty of it. Not so. The freshwater that we drink, bathe in, and irrigate our fields with, *is incredibly rare.* Only 3% of the world's water is freshwater, and 2/3 of that is tucked away in frozen glaciers, or otherwise unavailable for our use.

Water desalination might be the solution! Seawater desalination has the potential to reliably produce enough potable water to support large populations located near the coast.

Water desalination processes separate dissolved salts and other minerals from water. Feed water sources may include brackish, seawater, wells, surface (rivers and streams), wastewater, and industrial and process waters.

Membrane separation requires driving forces that include pressure (applied and vapor), electric potential and concentration to overcome natural osmotic pressures, to effectively force water through membrane processes. As such, the technology is energy intensive and research is continually evolving to improve efficiency and reduce energy consumption. Another important problem is the high contamination associated to this type of processes. The effluent from desalination plants is a multi-component waste, with multiple effects on water, sediment and marine organisms. It therefore affects the quality of the resource it depends on.

As a result, about 1.1 billion people worldwide lack access to water, and nearly 3 billion people find water scarce for at least 1 month of the year. Inadequate sanitation is also a problem for billions of people — they are exposed to diseases. Two million people, mostly children, die each year from diarrheal diseases alone.

Let us quote from the book by Solomon "water": "*Oil can be replaced — albeit painfully — by other fuel sources, while water cannot*". A paradigm shift is taking place. Most of the world's accessible water resources are built-out. With the absence of a new water breakthrough, which is not likely in the near future, countries around the world are about to go through an agitated and confused period like none before: "*the disparity between the Water-Haves and the Water Have-Nots will increase.*"

A breakthrough in reducing water consumption?

Simon Gepstein forgot for a while about the water-genetically-engineered plants he was taking care of and found that they could exist without water for three weeks and then come back to life with irrigation.

"Researchers at the Technion Faculty of Biology, led by Professor Simon Gepstein, Kinneret Academic College President, managed to genetically engineer plants that come into standstill "when they do not receive water for a while and come back and continue their development as renewable water supply, without damage to the plant caused. Professor Gepstein coincidental discovery came after he forgot for three weeks to water plants that have been engineered aiming to maximize their productive lives. It turned out that the same genetic changes make them also to consume less water. Is this a new hope?"

Epilogue

Historically, there have been water problems in the Middle East. In the Bible we read about conflicts and agreements around water, such as the one involving Abraham and Abimelech, the King of the Philistines, around the well called Beersheba.

Recently a research study has been published in the journal *Climatic Change*, which argues that a lack of water caused the collapse of the Assyrian empire (which stretches across Syria and Iraq today) about 2700 years ago. Archaeologists believe that a combination of severe drought and dense population in the capital of the empire of Nineveh caused thirst and hunger among the residents, which led to the swift collapse of the great empire of the ancient world. The lack of water and dense population still characterize the region, and although political and religious conflicts dominate today's conflicts agreements, we should never underestimate the weight of the water.

In the A-Disi region, in southern Jordan, a water conveyance system has been built with a huge investment. However, just before the system was operated, it turned out that the water there was radioactive and unfit for use. In Egypt, 10 million cubic meters of water from the Aswan Dam evaporates each year and is lost forever. In the Nile Delta in the Cairo area, water has turned into open sewer water because of pollution — millions of the people there pour their waste into it.

Turkey builds dams on the Tigris and Euphrates sources to stop the water from getting to the Syrians and Iraqis. Tehran has running water but an inefficient sewage system causes sewage to seep into groundwater, and some parts are flooding the city. Kuwait's groundwater is saturated with spilt oil from the first Gulf war and with salty water that was used to extinguish the fires that broke out. In addition, 8 dams and 30 water facilities were destroyed during the war, most of which did not return to work. Thirteen years later came the second Gulf War and this has resulted in 70% of the population in the war zones not getting a regular water supply since then.

With bad opening figures, Israel succeeded by using advanced technology, desalination and water recycling to become a kind of regional "water power". The gap between the discouraging water situation in Israel's neighboring countries and its vibrant situation is key to understanding the immediate and distant future of the Middle East. It is estimated that in 2050 the Middle East will have nearly 634 million people (double the current number), but with water sources are drying up, decreasing rainfall and global warming, water can be a major cause of conflict that might be even be bloodier than the ones we see today. But this may not necessarily be the case.

On the contrary water can be a source of stability, security, prosperity and peace in the "New Middle East", provided if we so desire. This vision brings the message of "water diplomacy", a hot new area in international relations. As part of water diplomacy, various parties are negotiating in terms of the distribution of water, based on the simplistic equation: — "an thirsty neighbor is a dangerous one". This approach is supported in numerous articles, research, expeditions and research group that claim: "do not miss a drop". Since we have not found a replacement for water that allows the existence of life, water has become an existential currency with which one can generate almost anything.

APPENDIX A

Photosynthesis

Photosynthesis is the most important biological process that takes place in the biosphere. Also, it is the only source of oxygen for aerobic organisms. Both eukaryotes and prokaryotes cells are able to carry out photosynthesis. Plant photosynthesis occurs in leaves and green stems within specialized cell structures called chloroplasts. One plant leaf is composed of tens of thousands of cells, and each cell contains 40 to 50 chloroplasts.

Apart from bacteria, the rest of photosynthetic organisms use water as an electron donor to reduce carbon dioxide according to the global equation:

$$nH_2O + nCO_2 + light \rightarrow (CH_2O)n + nO_2. \tag{A.1}$$

For the particular case $n = 6$, Eq. (A.1) leads to the formation of glucose ($C_6H_{12}O_6$) as the final product of photosynthesis. Such a reaction is exergonic and the energy required to carry it out comes from the light (photons). Therefore, the role of water in photosynthesis and consequently in preserving life on earth can hardly be oversized.

As we have shown in Chapter 4 (see Fig. 4.18), glucose is a carbohydrate that has a crucial role in life. The energy stored in its covalent bonds is the source of energy that powers cells in most organisms, from bacteria to humans.

Glucose is an energy source in most organisms from bacteria to humans. It is the human body's key source of energy, through aerobic respiration, and provides about 3.75 kcal of food energy per gram. The breakdown of carbohydrates (e.g., starch) yields mono and disaccharides, most of which is glucose. Through glycolysis and further reactions, glucose is oxidized to eventually form carbon dioxide and water, yielding energy sources, mostly in the form of adenosine triphosphate (ATP). The concentration of glucose in the blood is regulated through insulin reaction and other mechanisms.

Particularly, glucose is a primary source of energy for the brain, so its availability influences psychological processes. When glucose is low, psychological processes requiring mental effort (e.g., self-control, effortful decision-making, etc.) are impaired.

Sugar glucose is important because it is necessary for cellular respiration. Respiration is the typical process where the mitochondria of cells of organisms release chemical energy from sugar and other organic molecules through chemical oxidation. The global reaction, in a simplified way, can be written as:

$$Sugars + O_2 \rightarrow CO_2 + H_2O + released\ energy. \qquad (A.2)$$

This process occurs in both plants and animals.

During cellular respiration, the chemical energy in the glucose molecule is converted into a form that the plant can use for growth and reproduction. In the first step of respiration, called glycolysis, the glucose molecule is broken down into two smaller molecules called pyruvate, and a little energy is released in the form of ATP. This step in respiration does not require any oxygen and is therefore called anaerobic respiration. In the second step of respiration, the pyruvate molecules are rearranged and combined in a cycle. While the molecules are being rearranged in this cycle, carbon dioxide is produced and electrons are pulled off and passed into an electron transport system, which as in photosynthesis, generates a lot of ATP for the plant to use for growth and reproduction. This last step requires oxygen and therefore is called aerobic respiration. Thus, the final result of cellular respiration is that the plant consumes glucose and oxygen, and produces carbon dioxide, water and ATP energy molecules.

It is nice to see how photosynthesis and cellular respiration complement each other. During photosynthesis, the plant needs carbon dioxide and water, both of which are released into the air during respiration. And during respiration, the plant needs oxygen and glucose, which are both produced through photosynthesis.

We humans cannot produce glucose by ourselves (since we cannot produce photosynthesis in our bodies). We get this from plants although we do produce carbon dioxide that plants take in turn to produce oxygen that we breathe. Nice collaboration, calling us for taking care of plants!

Photosynthesis is a very complex process. For the sake of convenience and ease of understanding, plant biologists divide it into two stages. In the first stage — the light-dependent reaction — the chloroplast traps light energy and converts it into chemical energy contained in nicotinamide adenine dinucleotide phosphate (NADPH) and ATP, which produces oxygen. In the second stage — the dark reaction — carbon dioxide reduces, leading to glucose, with ATP providing the required energy. Therefore, the production of oxygen and reduction of carbon dioxide are two separated and differentiated processes.

Let us show some details of the two mentioned stages.

The very first evidence that the absorption of light energy (photons) induces an electron flow was provided by the British biochemist Robin Hill in 1937. Basically, an electron donor (H_2O) reacts with an electron acceptor, giving rise to the production of oxygen. This is the so-called Hill reaction. It was discovered that $NADP^+$ is the natural biological electron acceptor of green plants, and then, Hill reaction can be written as:

$$2H_2O + 2NADP^+ + Light \rightarrow 2NADPH + 2H^+ + O_2. \quad (A.3)$$

When a chlorophyll molecule is excited by the absorption of a photon, some electrons occupy higher energy levels and then leave the chlorophyll molecule and incorporate into an electron transport chain, giving rise to NADPH (the reduced form of $NADP^+$). Replacement of these electrons, to ensure continuity in the process, is carried out by H_2O according to the reaction:

$$2H_2O + Light \rightarrow 4H^+ + 4e^- + O_2. \quad (A.4)$$

Further investigations showed that the electron flow is accompanied by the formation of ATP and H_2O, from ADP (adenosine diphosphate) and P_i (inorganic phosphate), using the energy of sunlight. The reaction is known as photophosphorylation.

So the global reaction can be written as:

$$H_2O + NADP^+ + 2P_i + 2ADP + Light$$
$$\rightarrow NADPH + H^+ + 1/2O_2 + 2ATP + 2H_2O, \quad (A.5)$$

where the water molecule on the left-hand side of Eq. (A.5) is the electron donor, while the two water molecules on the right-hand side come from the formation of ATP from ADP and P_i.

Regarding the light independent stage of photosynthesis, the American biochemist Melvin Calvin was awarded the 1961 Nobel Prize in Chemistry for his studies on the conversion of carbon dioxide into carbohydrates (Calvin cycle). The very complex set of equations involved can be summarized into the global reaction,

$$6CO_2 + 12NADPH + 18ATP + 12H^+$$
$$\rightarrow C_6H_{12}O_6 + 12NADP^+ + 18P_i + 18ADP. \quad (A.6)$$

The reduction of six molecules of carbon dioxide implies 24 electrons,

$$6CO_2 + 24H^+ + 24e^- \rightarrow C_6H_{12}O_6 + 6H_2O. \qquad (A.7)$$

If, according to the experimental observation, for each pair of electrons, 2 phosphorylation reactions take place, 24 ATP molecules and 12 NADPH molecules would be produced, bearing in mind Eq. (A.5). Therefore, the light dependent stage of photosynthesis produces more ATP than that required to reduce six molecules of CO_2 during dark reactions. Of course, plants need ATP (energy deposit) for many other purposes!

To sum up, solar energy is converted into chemical energy during photosynthesis. This chemical energy is stored in the form of glucose. Carbon dioxide, water and sunlight are used to produce glucose and oxygen, according to the reaction:

$$6CO_2 + 6H_2O + Light \rightarrow C_6H_{12}O_6 + 6O_2. \qquad (A.8)$$

It was the evolution of this system, some two billion years ago that began the process by which oxygen is released into our atmosphere, making possible the evolution of life, as we know it in this world. In this process the role of water as an electron donor in photosynthesis is of utmost importance.

Solvation of Charged Particles

As has been remarked in several parts of the book, one of the most relevant roles played by water in Nature is that of a solvent. We have seen that the polar structure of solvent, with partial charges on the constituent atoms, is responsible for the stabilizing interactions established between water and the solvated substances.

In this section, we will present and analyze in some detail one of the simplest theoretical models, introduced in 1920 by the mathematician and physicist Max Born, who was awarded the Noble Prize in Physics in 1954 for studying solvation effects in a quantitative manner. The model, with slight variations, is useful to estimate the solvation energy, even till today.

B.1. Some Elementary Electrostatics

It is well-known that the electric potential created by a point charge $q(\mathbf{r}_1)$ on a given point in the space ($\mathbf{r}_2$) is given by

$$\phi_1(\mathbf{r}_2) = \frac{1}{4\pi\varepsilon}\frac{q(\mathbf{r}_1)}{d_{12}}, \tag{B.1}$$

where ε is the permittivity of the medium (the permittivity is a measure of how much the molecules oppose the external electric field. See Appendix D) and d_{12} is the distance between points 1 and 2.

Let us put a point charge $q(\mathbf{r}_2)$ on position 2. The electrostatic potential energy will be given by:

$$q(\mathbf{r}_2) \cdot \phi_1(\mathbf{r}_2) = q(\mathbf{r}_1) \cdot q(\mathbf{r}_2)/4\pi\varepsilon d_{12}. \qquad (B.2)$$

If we compute in a similar way the electrostatic potential energy of a point charge $q(\mathbf{r}_1)$ on point $\mathbf{r}_2$ where the electrostatic potential created by a charge $q(\mathbf{r}_2)$ is $\phi_2(\mathbf{r}_1)$, we obtain, of course, an identical result:

$$q(\mathbf{r}_1) \cdot \phi_2(\mathbf{r}_1) = q(\mathbf{r}_1) \cdot q(\mathbf{r}_2)/4\pi\varepsilon d_{12}, \qquad (B.3)$$

because in both cases we are computing the same thing, namely, the potential interaction energy of two point charges $q(\mathbf{r}_1)$ and $q(\mathbf{r}_2)$. Therefore, in order to compute the potential energy U we can employ the formula:

$$U = \frac{1}{2} \sum_{i} \sum_{j \neq i}^{2} q(\mathbf{r}_i)\phi_j(\mathbf{r}_i), \qquad (B.4)$$

where the factor $1/2$ avoids the problem of counting twice the identical contributions mentioned in the previous paragraph.

Equation (B.4) can be extended to the general case, where we consider a N-particle system by means of the equation

$$U = \frac{1}{2} \sum_{i}^{N} q(\mathbf{r}_i)\phi(\mathbf{r}_i), \qquad (B.5)$$

where $\phi(\mathbf{r}_i)$ represents the potential on the position $\mathbf{r}_1$ created by all particles but $q(\mathbf{r}_i)$.

If instead point charges, we consider a charge density $\rho(\mathbf{r}_1)$ (charge per volume unit), so we pass from the sum to the integral

$$U = \frac{1}{2} \int \rho(\mathbf{r}_i)\phi(\mathbf{r}_i)d\tau, \qquad (B.6)$$

where $d\tau$ is the volume element.

U in Eq. (B.6) represents the work required to create a charge distribution and according to thermodynamics, it is nothing else than the change in Gibbs free energy at constant temperature and pressure, ΔG.

We have stressed in previous chapters that ions become stabilized in solution by the interaction with solvent (water). How can we estimate the solvation energy of an ion in water solution? Let us assume a rather simplified model to represent an ion: A charged sphere of radium R immersed in water represented by a continuum medium, characterized by a dielectric constant ε (see Fig. B.1). According to electrostatics, the charge of such a sphere spreads over the surface. From Gauss' theorem we know that the field inside the sphere is null, $\mathbf{E} = 0$, and the potential is constant, with a value $q/4\pi\varepsilon R$.

The charge density (charge/surface) is $\rho(\mathbf{r}) = q/4\pi R^2$ and the potential $\phi(\mathbf{r}_i) = q/4\pi\varepsilon R$. Then,

$$\Delta G = U = \frac{1}{2}\int \rho(S)\phi(\mathbf{r}_i)dS = \frac{q^2}{8\pi\varepsilon R}. \qquad \text{(B.7)}$$

Now we are in the position to compute the change in Gibbs free energy for the solvation of an ion. We adopt the rather simple model proposed by Born and depicted in Fig. B.2.

According to thermodynamics, the solvation energy of an ion (ΔG_{Born}) can be estimated through the alternative pathway $\Delta G_a + \Delta G_b + \Delta G_c$, where ΔG_a stands for the energy required to

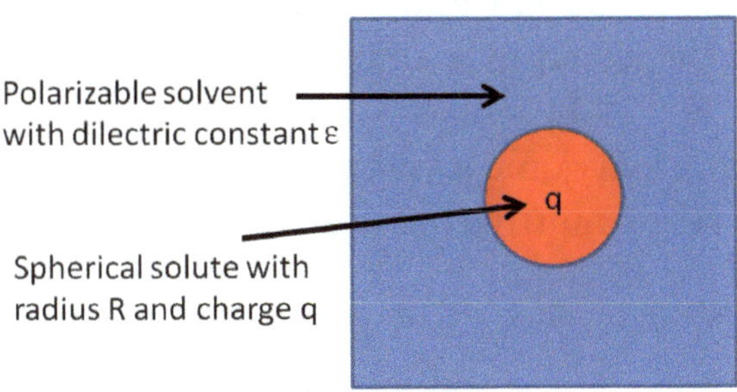

Fig. B.1.   Simplified model for an ion.

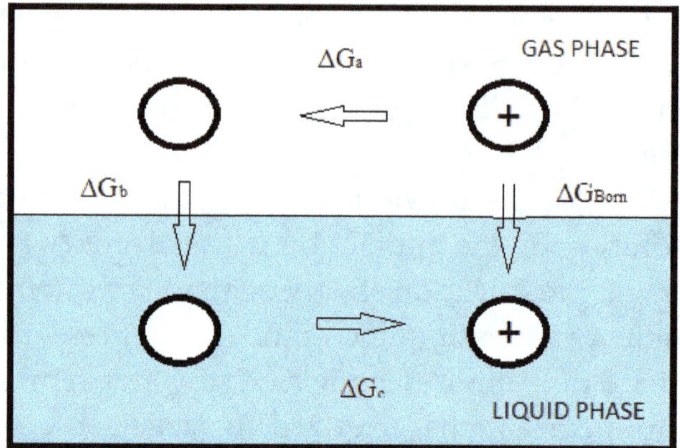

Fig. B.2.   Born model for solvation.

discharge the ion in "gas phase" (vacuum), ΔG_b is the energy to transfer the discharged ion from vacuum to the "liquid phase" (water solution) and ΔG_c is the energy to recharge the ion in water. It is important to note that in this model, we are simplifying the solvent as a continuum electric field characterized by a dielectric constant. The solute–solvent interaction is therefore the interaction between the ionic charge and the solvent electric field. We do not include in the model the ion–ion interactions, so we are dealing with what is called an ideally dilute solution.

ΔG_b represents the interaction energy of a neutral molecule with the solvent (water), and such a contribution is expected to be much smaller than ΔG_a and ΔG_c, which involve interactions between the charged ion and water, a polar molecule. Thus, ΔG_b can be neglected in a simple model like the one we are introducing. For ΔG_a and ΔG_c we employ Eq. (B.7) with ε_0 (vacuum permittivity) and ε_c (water permittivity) $= \varepsilon_0 \cdot \varepsilon_r$, where ε_r is the relative permittivity of water.

Then,

$$\Delta G_a = -\frac{q^2}{8\pi\varepsilon_0 R}; \quad \Delta G_c = \frac{q^2}{8\pi\varepsilon_0\varepsilon_r R}. \tag{B.8}$$

The negative sign in ΔG_a arises from the fact that the system (charged ion) must carry out work in order to get discharged, in a similar way that a compressed gas works on the surroundings to expand itself.

Then, according to the convention established in thermodynamics: $w_a < 0$ and $\Delta G_a < 0$, where w_a is the work done by the system in order to get discharged.

In the case of ΔG_c, the surroundings must work on the system in order to charge it ($w_c > 0$ and $\Delta G_c > 0$), in a similar way that the surroundings work on a gas to compress it in a piston.

Finally, we get,

$$\Delta G_{\text{Born}} = -\frac{1}{2}\left(1 - \frac{1}{\varepsilon_r}\right)\frac{q^2}{4\pi\varepsilon_0 R}, \qquad (\text{B.9})$$

which allows us to estimate in a rather simple way the solvation energy of an ion ($q = Ze$, where Z is the charge and e is the electron charge) in water ($\varepsilon_r = 79.54$).

How does the Born model work? Surprisingly, the model works reasonably well, bearing in mind its extreme simplicity. For example, experimental data shows that the chlorine ion (Cl^-) is about $61\,kJ/mol$ more stable in water than the iodine ion (I^-). A simple calculation using Eq. (B.9) with the appropriate units and the ionic radii $R(Cl^-) = 181\,pm$ and $R(I^-) = 220\,pm$ (pm = picometer = $10^{-12}\,m$) gives a value of $-67\,kJ/mol$.

Equation (B.9) can be easily generalized to account for electrostatic contributions other than those arising from charges (dipoles, quadrupoles, octupoles, ...) and continues working fairly well. Of course, Eq. (B.9) neglects a number of contributions that on occasions can be non-negligible. For example, Eq. (B.9) does not consider the non-electrostatic components of the solvation Gibbs free energy (see ΔG_b above). Other cavities with non-spherical shape for the ion are not contemplated in this simple model (complex molecules produce non-spherical cavities). Another important limitation is the total exclusion of hydrophobic interactions, which have been shown to be crucial for a number of biological systems, as remarked in previous chapters.

On the other hand, it is clear that modeling the solvent as a continuum system is a very crude approach. During the last decades, more appropriate (realistic) models including all the physically relevant contributions are being employed. Particularly promising are those discrete models that consider explicitly the water molecules in the calculations. We will not provide more details here as our purpose is to provide the reader with a basic idea of how the role of water as a solvent can be described, using quantitative physical models.

Tides

Tides are the periodic vertical movement of water on the Earth's surface caused by the combined effects of the gravitational forces exerted by the Moon and the Sun, and the rotation of the Earth, giving rise to the sea advancing and retreating once or twice a day, which we can observe on our beaches. The Ancient Greeks and Romans were not particularly concerned with the tides at all, since they are almost imperceptible in the Mediterranean. Caesar lost war galleys on the English shores due to his ignorance about tides.

Nevertheless, Eratosthenes and Aristotle were aware of the relation between the tides and the Moon. Pytheas, the Greek geographer and explorer from Massalia, was the first to associate the tides to the phases of the Moon. Pliny the Elder describes in detail the phenomenon of tides in his *Naturalis Historia* and relates it with the Moon and the Sun.

It appears that Kepler realized the cause of the tides but the subject was first investigated mathematically by Newton in 1687, when he applied the gravitational law to the explanation of tides. Newton supposed that the ocean covers the whole Earth and that the water remains always in a state of equilibrium under the combined influence of the Earth, Moon and Sun.

According to Newton's law of gravity, a body of mass M exerts a gravitational attraction on another mass m given by:

$$\mathbf{F} = -G\frac{M \cdot m}{r^2}\mathbf{r}, \qquad (C.1)$$

where **r** is a unit vector connecting both bodies that are at the distance r.

Let us now compare the gravitational effect of the Sun and the Moon on the Earth: The mass of the Sun is about 27 million times greater than that of the Moon, and the distance of the Sun to the Earth is about 390 times the distance of the Moon to the Earth. With this data, Eq. (C.1) tells us that the gravitational attraction of the Sun is about 178 times greater than that of the gravitational attraction of the Moon. Is then the Sun more effective than the Moon in producing tides? The answer is, of course, no!

Figure C.1 helps us to explain this apparent contradiction. Let us consider point A in this figure. The tidal force ($\mathbf{f}_A$) of a mass of water m is defined as $\mathbf{F}_A - \mathbf{F}_T$,

$$\mathbf{f}_A = \mathbf{F}_A - \mathbf{F}_T = -GMm \left\{ \frac{1}{(r-R)^2} - \frac{1}{r^2} \right\} \approx GMm\frac{2R}{r^3}\mathbf{i},$$

$$(C.2)$$

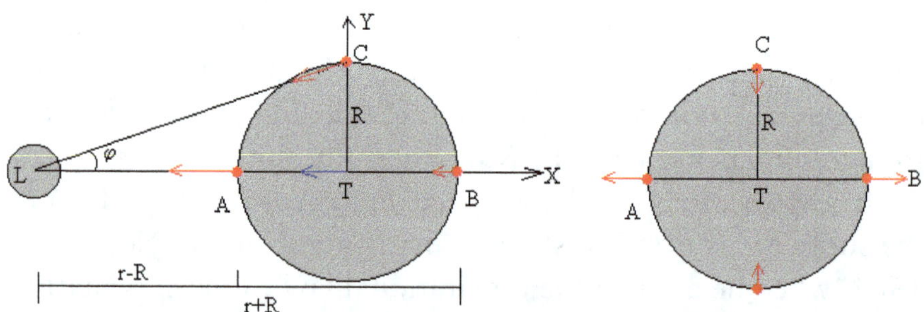

Fig. C.1.   A, B and C represent points on the Earth's surface. If we consider that the ocean covers the entire Earth, water (a mass m of water) on A, B and C will experience gravitational attraction towards the Moon: $\mathbf{F}_A$, $\mathbf{F}_B$ and $\mathbf{F}_C$ (in red), respectively. $\mathbf{F}_T$ (in blue) corresponds to the gravitational attraction towards the Moon if the mass of water was at T, the center of the Earth. The scheme on the right shows the tidal forces (the difference between the red and the blue vectors).

where r stands for the distance between the center of the Earth and the center of the Moon, and R is the radius of the Earth. M is the mass of the Moon, and $\mathbf{i}$ is a unit vector in the $\mathbf{x}$ direction.

In the same way, it can be easily shown that:

$$\mathbf{f_B} \approx GMm\frac{2R}{r^3}\mathbf{i}; \quad \mathbf{f_C} \approx -GMm\frac{2R}{r^3}\mathbf{j}, \tag{C.3}$$

where j is a unit vector in the y direction.

Equations (C.2) and (C.3) make clear two fundamental points: (a) The tidal force varies with the third power of r, and not with the second power of r as in Eq. (C.1). Therefore, the tide raising forces of the Sun are approximately $178/390 = 0.46$ times than that of the Moon. Yes, according to our experience, is not the Sun, but the Moon which exerts a greater influence on the tides, and (b) the gravitational force the Moon exerts on any body on the surface of the Earth is much smaller than the gravitational force of the Sun. However, because the Moon is much closer to the Earth than the Sun, the lunar gravitational field across the Earth is considerably greater than that of the solar field. As a result, Moon-induced tides are more than twice as great as Sun-induced tides.

$\mathbf{F_A}$ and $\mathbf{F_B}$ are both directed vertically upward from the Earth towards the Moon. Against intuition, the tidal forces on points A and B (equatorial antipodes) produce similar effects, namely, water rise (flood) in both cases (see Fig. C.2). One would be tempted to predict a water fall on B.

With a little bit more complex calculations, one can show that the sea level rise can be estimated from:

$$h \approx \frac{M}{2M_T}\left(\frac{R}{r}\right)^3 R(3\cos^2\theta - 1), \tag{C.4}$$

where M is the mass of the perturbing body (the Moon or Sun), M_T is the Earth's mass and θ is the LTP (P $=$ A, B, C) angle in

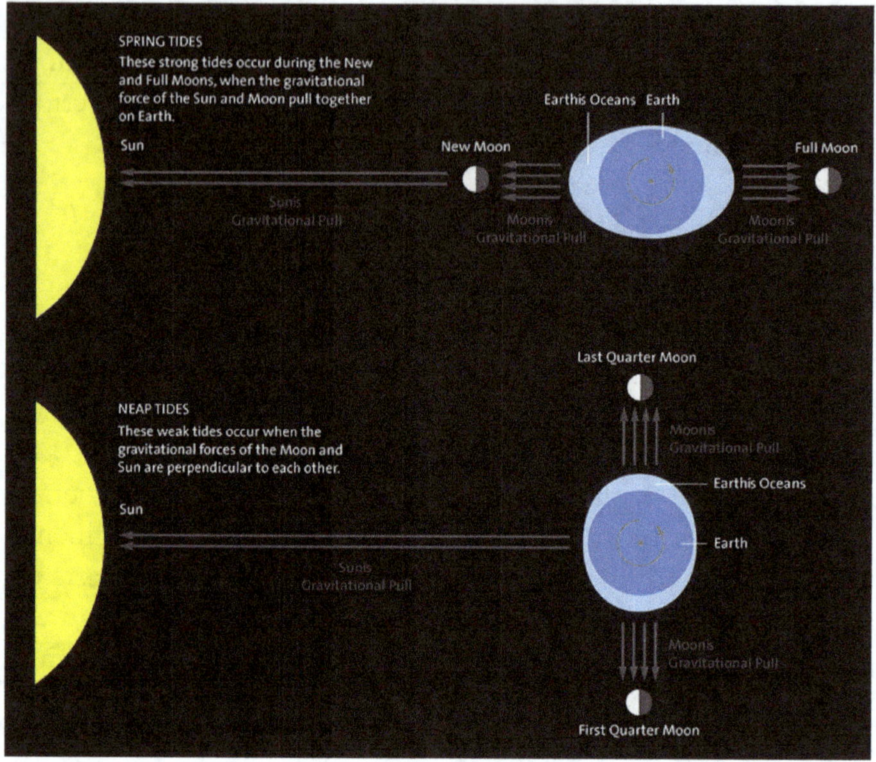

Fig. C.2. Effects of the Sun and the Moon on the Earth's oceans: The formation of tides.

Fig. C.2. The maximum value is for $\theta = 0, \pi$ (points A and B) and the minimum value corresponds to $\pi = \pi/2$ (point C).

It can be seen that $|h_{\max}| = 2|h_{\min}|$ and the difference between the maximum water highs in high and low tides is:

$$\Delta h_{\max} \approx \frac{3M}{2M_T} \left(\frac{R}{r}\right)^3 R. \tag{C.5}$$

The reader should not forget that previous equations and considerations are only valid for an oversimplified model called the Equilibrium Tide Theory, where we consider an imaginary Earth that is totally covered in water and has no land masses (e.g., continents, islands).

There are several important factors that modify the movement of water in real tides situations: Land masses obviously impede and deflect movement of water on the Earth's surface, friction retards the movement of water particles across the Earth's surface, lunar and terrestrial orbits change in cycles of months, years and even longer periods, the Earth's and Moon's orbits are complicated eccentric ellipses, and the Earth's declination (23°27′ off the vertical) causes important effects on tides, etc. The study of tides is a very complicated subject but with an enormous practical impact. The so-called tide tables, elaborated after laborious calculations, provide tidal prediction and show the daily times and heights of high water and low water for any particular location. They are calculated and published for standard ports (major ports) and for a period of one year.

Why do we need tidal predictions? Knowledge of the times and heights of tides and the speed and direction of tidal streams is important to a variety of people. These include: (a) Naval forces: they would not be able to defend territorial waters or attack enemy coasts without accurate charts. (We already mentioned that Caesar, a native of the Mediterranean, where the tides are practically undetectable, was taken by surprise by high British tides.) (b) Navigators: particularly in estuarine and coastal waters and the approaches to harbors, (c) harbor and coastal engineers: in the construction of harbor works, bridges, locks and dykes, and (d) the public: to know when to go fishing, sailing, cross rivers on hiking trails, etc.

Dielectric Properties of Water

The dielectric properties, or *permittivity*, is one of the factors that determines how a material interacts with an applied electromagnetic field.

Electric fields cause current to flow through a conductor in which electrons are free to move. Well, not entirely free since they transfer energy in the form of heat to the environment. But there is no limitation to their movement. In the so-called dielectrics, material electrons are bounded to the molecules and atoms. An electric field causes limited motion of the electrons that are bound to the atom or molecule and polarizes them. Positive charges are displaced along the field, while the negative charges are displaced in a direction opposite to that of the field, thus causing the creation of opposite charges on either side of the molecule

Field polarizes the electron cloud, converting a neutral atom into a "smeared" dipole (see Fig. D.1). In that case, the so-called "dipole moment" measures the average of internal separation of opposite charges. Mathematically, one can get an elegant correction for any electron distribution, whatever it is. This is the power of mathematics, which we save you.

If the molecule contains ionic bonds, then the field tends to stretch the lengths of these bonds. Stretching depends on the "flexibility", so to speak, of the molecule; a property called *polarizability*. The polarizability *determines the response of the molecule to an external electric field in terms of the degree of*

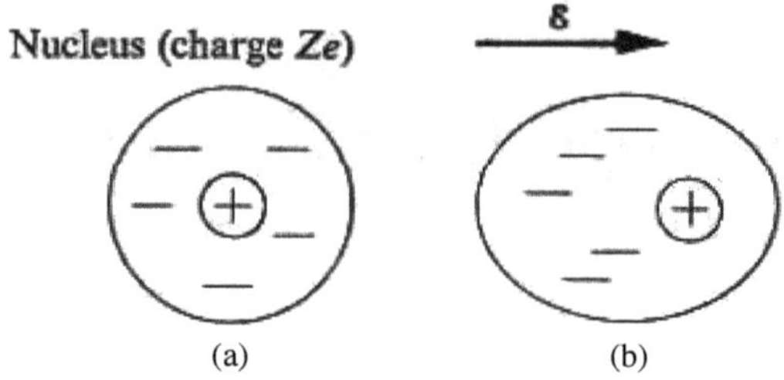

Fig. D.1.   Electron polarization of an atom.

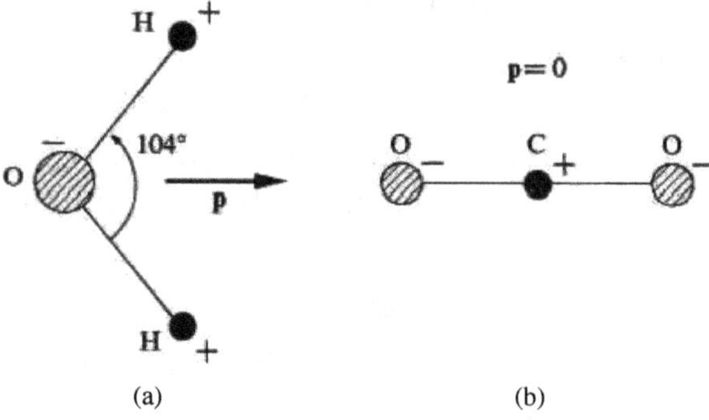

Fig. D.2.   Permanent polarization.

stretching: The dipole moment increases under the effect of a static electric field (or its component, acting along the axis connecting the ions).

A different kind of polarization of much importance to the response of water to an electric field is the orientation polarization. The individual molecules in a dipolar substance are permanent dipoles, even in the absence of an electric field. We speak then of a *dipolar* molecule.

The dipole moments of the two OH "arms" of the H_2O molecule add to give a net dipole moment. In the CO_2 molecule (see Fig. D.2), the dipole moments of the two CO arms are of opposite directions — they cancel each other and their net dipole moment is zero.

Measuring dipole moments teaches a lot about the structure of the molecule.

In bulk water, we are dealing with an enormous number of molecules. Their *net total polarization vanishes* in the absence of an external field since the molecular dipoles are randomly oriented (see Fig. D.3), thus resulting in a complete cancellation of the polarization. When a field is applied, the molecular dipoles tend to align with the field, not entirely but the orientation is changed in favor of the field direction and this *results in a net non-vanishing average polarization*, which in the absence of a field is, on average, zero.

Molecular rotation due to the field is however inhibited to some degree and slowed down due to friction with neighboring molecules depending on viscosity.

We are interested in the observed response of water to these three mechanisms of reaction to an electric field and relation to water properties.

Let us focus on a static electric field. In a vacuum, an electric charge on a parallel condenser (see Fig. D.4) gives rise to a potential difference, V. The introduction of water into the space

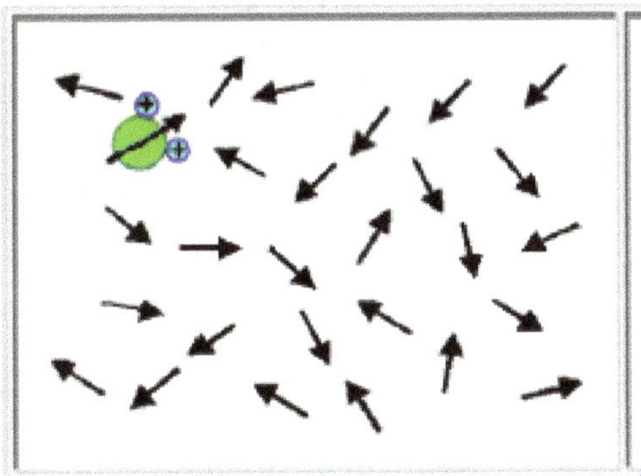

Fig. D.3.   Randomly oriented dipolar molecules. Net polarization vanishes. Dipolar molecules tend to align with an external field but rotational movements are inhibited due to friction. A net polarization is created.

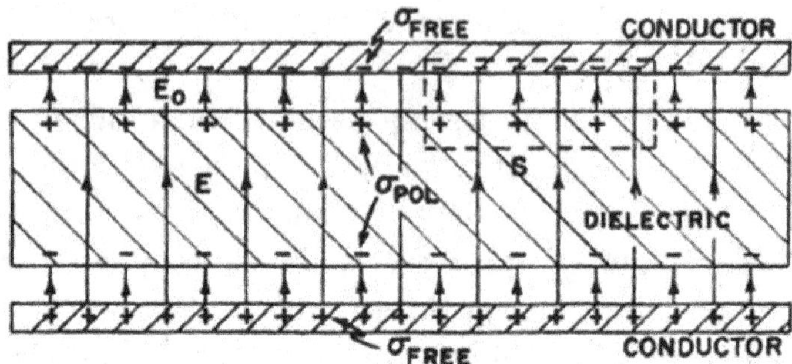

Fig. D.4.   The effect of a dielectric body inserted in between charged plates.

between the plates causes the reduction of V by a factor of 81: This is the *static dielectric constant*. Take a look on the double plate condenser in Fig. D.4 where a solid dielectric slab fills the space between the two plates. The dipoles — whether permanent or polarized by the electric field between the plates — are aligned with the direction of the field. Inside the slab the effect of the charges cancel each other. However, the opposite charges on the two surfaces of the slab *weaken the effect* of the opposite charge on each of the plates, thus *reducing the electric field in between the plates.*

Let us be familiar with the basic nomenclature and equations in relation to the above qualitative description: If σ_{free} is the free charge density of opposite signs on each plate in Fig. D.4, we know from elementary electrostatics that if the space between the two plates is empty, the value of the electric field in between the two plates is:

$$E = 4\pi\sigma. \tag{D.1}$$

If we insert between the two plates a dielectric material containing dipoles that orient themselves in the perpendicular direction to the plane of the plates (might be initially electrically neutral atoms in which electrons were shifted toward the positive plate, thus creating small dipoles) an additional electric field is created in the opposite direction to the original one (see Fig. D.4), and E is reduced to the value:

$$E = \frac{4\pi\sigma}{\varepsilon}. \tag{D.2}$$

The true charge density σ on each plate is unchanged, but the new effective charge density on each side of the system is due to the charges on the *surfaces of the dielectric* that oppose the true initial charge density on the plates and the total surface density is reduced by the amount:

$$P = \sigma\left(1 - \frac{1}{\varepsilon}\right). \tag{D.3}$$

One can say that E in Eq. (D.1) is replaced by a new field quantity, which is described by the "true" charges on the plates:

$$D = 4\pi\sigma = \epsilon E = E + 4\pi P, \tag{D.4}$$

where ϵ is the *dielectric constant* or *relative permittivity*.

Let us come back to water. There is a similar situation to the inserted dielectric material between two charged plates if water molecules are in between them. Water molecules that float towards the charged plates orient their positive side to the negative plate, and vice versa, reducing the effective charge on each side and lowering the electric field.

The average electric field between the plates is reduced by a factor that is the dielectric constant. The same reduction applies to the electric field between two ions solved in water (Na^+ and Cl^-) when NaCl is dissolved in water. The positive parts of the water molecule are attracted to the negative Cl^- ion and surround it just as they were attracted to the negative plate. The negative field induced by the chlorine atom is reduced. The negative parts of the water molecule behave in the same manner around the positive Na^+.

References and Suggested Reading

Anfinsen, C. B., Haber, E., Sela, M., White, F. H. Jr. (1961) *Science* **181**: 223.

Anfinsen, C. B., Haber, E., Sela, M., White, F. H. Jr. (1961) The kinetics of formation of native ribonuclease during oxidation of the reduced polypeptide chain. *Proc. Natl. Acad. Sci. USA* 1309.

Anfinsen, C. B. (1973) Principles that govern the folding of protein chains. *Science* **181**: 223.

Antonini, E., Brunori, M. (1971) *Hemoglobin and Myglobin in their Reactions with Ligands*. North-Holland, Amsterdam.

Baldwin, R. L. (1995) The nature of protein folding pathways: The classical versus the new view. *J. Biomol. NMR* **5**: 103.

Ball, P. (1999) H_2O, *A Biography of Water*. Weidenfeld and Nicholson London.

Ben-Naim, A. (1974) *Water and Aqueous Solutions, Introduction to Molecular Theory*. Plenum Press, New York.

Ben-Naim, A. (1980) *Hydrophobic Interactions*. Plenum Press, New York.

Ben-Naim, A. (1990) *Biopolymers* **29**: 567.

Ben-Naim, A. (1992) *Statistical Thermodynamics for Chemists and Biochemists*. Plenum Press, New York.

Ben-Naim, A. (2001) *Cooperativity and Regulation in Biochemical Processes*. Kluwer Academic/Plenum Publishers, New York.

Ben-Naim, A. (2009) *Molecular Theory of Water and Aqueous Solutions, Part I: Understanding Water*. World Scientific, Singapore.

Ben-Naim, A. (2009) *Molecular Theory of Water and Aqueous Solutions: Understanding Water*. World Scientific.

Ben-Naim, A. (2011) *Molecular Theory of Water and Aqueous Solutions, Part II: The Role of Water in Protein Folding Self Assembly and Molecular Recognition*. World Scientific, Singapore.

Ben-Naim, A. (2016) *Entropy, The Truth, the Whole Truth and Nothing but the Truth*. World Scientific, Singapore.

Benedict, W. S., Gailar, N. and Plyler, E. K. (1956) *J. Chem. Phys.* **24**: 1139.

Chang, R. (2005) *Physical Chemistry for the Biosciences*. University Science Books, United States.

Cohn, E. J., Edsall, J. T. (1943) In *Proteins, Amino Acids and Peptides as Ions and Dipolar Ions*. Editors: Cohn, E. J., Edsall, J. T. Reinhold, New York, Chapter 9.

Corey, R. B., Pauling, L. (1953) *Proc. R. Soc. London Ser. B* **141**: 10.

Creighton, T. E. (1984) *Proteins, Structure and Molecular Principles.* Freeman, New York.

Creighton, T. E. (1993) *Proteins, Structure and Molecular Principles*, 2nd edition. W. H. Freeman and Company, New York.

Dorsey, N. E. (1940) *Properties of Ordinary Water-Substance.* Reinhold, New York.

Edsall, J. T. and Wyman, J. (1958) *Biophysical Chemistry, Volume 1.* Academic Press, Inc., New York.

Eisenberg, D. and Kauzmann, W. (1969) *The Structure and Properties of Water.* Oxford University Press, Oxford, UK.

Franks, F. (1982) *Water: A Comprehensive Treatise* (F. Franks, ed), Volume 7, *Water and Aqueous Solutions at Subzero Temperatures*, Plenum Press, New York.

Franks, F. (2000) *Water, a matrix of life*, 2nd edition. The Royal Society of Chemistry.

Haber, E., Anfinsen, C. B. (1961) *J. Biol. Chem.* **236**: 422.

Henderson, L. J. (1913) *The Fitness of the Environment. An Inquiry into the Biological Significance of the Properties of Matter.* McMillan Co., New York.

Herzberg, G. (1950), *Molecular Spectra and Molecular Structure*, Van Nostrand, New York.

Kauzman, W. (1959) *Adv. Protein Chem.* **14**: 1.

Kavanau, J. L. (1964) *Water and Solute-Water Interactions.* Holden Day, San Francisco, California.

Levinthal, C. (1968) Are there pathways for protein folding? *J. Chim. Phys.* **65**: 44.

Levinthal, C. (1969) In *Mossbauer Spectroscopy in Biological Systems.*

Pauling, L. (1960) *The Nature of Chemical Bond*, 3rd edition. Cornell University Press, Ithaca, New York.

Pauling, L. (1960) *The Nature of the Chemical Bond*, 3rd edition. Cornell University Press, New York.

Perutz, M. F. (1990) *Mechanisms of Cooperativity and Allosteric Regulation in Proteins.* Cambridge University Press.

Robinson, G. W., Zhu, S. B., Singh, S., and Evans, M. W. (1996) *Water in Biology, Chemistry and Physics.* World Scientific, Singapore.

Samoilov, O. Ya (1957) *Structure of Aqueous Electrolyte Solutions and Hydration of Ions* (translated by D.J.G. Ives), Consultants Bureau, New York (1965).

Stryer, L. (1988) *Biochemistry*, 3rd edition. W. H. Freeman and Company, New York.

Tanford, C. (1980) *The Hydrophobic Effect.* Wiley, New York.

Tanford, C., Reynolds, J. (2001) *Nature's Robots, A History of Proteins.* Oxford University Press, Oxford.

Voet, D. J., Voet, J. G., Pratt, C. W. (2008) *Principles of Biochemistry*, 3rd edition. John Wiley and Sons, New Jersey.

Volkenstein, M. V. (1977) *Molecular Biophysics.* Academic Press, New York.

Wang, H., Ben-Naim, A. (1997) *J. Phys. Chem. B* **101**: 1077, 1086.

Whitford, D. (2005) *Protein, Structure and Function.* John Wiley and Sons, United Kingdom.

Index